£37.40
.int

KT-559-371

Robot Sensors

International Trends in Manufacturing
Technology

ROBOT SENSORS
Vol. 1 – Vision

Edited by
Professor Alan Pugh

IFS (Publications) Ltd, UK

Springer-Verlag
Berlin Heidelberg New York Tokyo
1986

British Library Cataloguing in Publication Data

Robot sensors.—(International trends in manufacturing technology)
    Vol. 1 : Vision
    1. Robots   2. Pattern recognition systems
    I. Pugh, A.   II. Series
    629.8'92   TJ211

ISBN 0-948507-01-2 IFS (Publications) Ltd
ISBN 3-540-16125-2 Springer-Verlag Berlin Heidelberg New York Tokyo
ISBN 0-387-16125-2 Springer-Verlag New York Heidelberg Berlin Tokyo

© 1986 **IFS (Publications) Ltd,** 35-39 High Street, Kempston,
Bedford MK42 7BT, UK
and **Springer-Verlag** Berlin Heidelberg New York Tokyo

This work is protected by copyright. The rights covered by this are reserved, in particular those of translating, reprinting, radio broadcasting, reproduction by photo-mechanical or similar means as well as the storage and evaluation in data processing installations even if only extracts are used. Should individual copies for commercial purposes be made with written consent of the publishers then a remittance shall be given to the publishers in accordance with §54, Para 2, of the copyright law. The publishers will provide information on the amount of this remittance.

Phototypeset by Wagstaffs Typeshuttle, Henlow, Bedfordshire
Printed and bound by Short Run Press Ltd, Exeter

85 011339

International Trends in Manufacturing Technology

The advent of microprocessor controls and robotics is rapidly changing the face of manufacturing throughout the world. Large and small companies alike are adopting these new methods to improve the efficiency of their operations. Researchers are constantly probing to provide even more advanced technologies suitable for application to manufacturing. In response to these advances IFS (Publications) Ltd is publishing a series of books on topics that highlight the developments taking place in manufacturing technology. The series aims to be informative and educational.

Subjects already covered in the series include:

Robot Vision, Programmable Assembly, Robot Safety, Robotic Assembly, and Flexible Manufacturing Systems.

Other subjects to be covered include:

Electronics Assembly, Robotic Welding, Automated Guided Vehicles, Robot Grippers, Education and Training for Robotics and Automation, Human Factors in Advanced Manufacturing Systems, Simulation in Manufacturing, Artificial Intelligence.

The series is intended for manufacturing managers, production engineers and those working on research into advance manufacturing methods. Each book will be published in hard cover and will be edited by a specialist in the particular fields.

This, the sixth in the series – Robot Sensors, Vol. 1 – is under the editorship of Professor Alan Pugh of the University of Hull, UK. The series editors are: Michael Innes, John Mortimer, Brian Rooks, Jack Hollingum and Anna Kochan.

Finally, I express my gratitude to the authors whose works appear in this publication.

John Mortimer,
Managing Director,
IFS (Publications) Ltd

Acknowledgements

IFS (Publications) Ltd wishes to express its acknowledgement and appreciation to the following publishers/organisations for granting permission to use some of the papers reprinted within this book.

IFS (Conferences) Ltd
35-39 High Street
Kempston
Bedford MK42 7BT
England

John Wiley and Sons, Inc.
605 Third Avenue
New York, NY 10158
USA

GEC Research Laboratories
Marconi Research Centre
West Hanningfield Road
Chelmsford
Essex CM2 8HN
England

Coventry Lanchester Polytechnic
Priory Street
Coventry CV1 5FB
England

Society of Manufacturing Engineers
One SME Drive
PO Box 930
Dearborn, MI 48121
USA

SRI International
333 Ravenswood Avenue
Menlo Park, CA 94025
USA

Japan Industrial Robot Association
c/o Kikaishinko Building
3-5-8 Shiba Koen
Minato-ku
Tokyo
Japan

The Institute of Physics
Techno House
Redcliffe Way
Bristol BS1 6NX
England

National Research Council of Canada
Montreal Road
Ottawa
Ontario K1A OR8
Canada

Plenum Press
227 West 17th Street
New York, NY 10011
USA

National Science Foundation
1800 G Street North-west
Washington, DC
USA

General Motors Research Laboratories
Warren, MI 48090-9055
USA

Contents

1. Reviews

2. Special Vision Sensors

3. Fibre-Optic Sensors

Preface

Robot sensors have been the subject of much research in recent years; indeed two volumes are required to present adequately the quantity of individually published work in this area. This first volume covers aspects of 'vision'. It is naively assumed that sensors for robot vision applications are in plentiful supply. This implies that cameras, designed for television applications, are acceptable for use in robotic applications but experience shows that these products need extensive modification to tolerate the rugged environment of a robotic workcell. In fact this book on vision sensors makes little reference to conventional solid-state commercial television cameras. Indeed, it is not an exaggeration to claim that there is a crisis surrounding the robot sensor market even though much published research relies on the assumption that 'the sensor' is forthcoming. When attempts are made to apply the results of research in industry the realisation that the sensor needed is just not available completely defeats the objective.

Although, curiously, little has been written on the subject of vision sensors, this book is representative of the most notable developments in the field. Readers might be surprised to find that Volume 2, dealing with 'non-vision sensors', reveals much more activity in terms of research and some results in terms of commercial exploitation. *The two volumes should be used together* to provide a comprehensive collection of papers representing both historical developments and recent ideas from authors who have published in this area.

In this volume, the first section includes a review of the sensor issue presented by the editor earlier this year. This is supported by a detailed paper evaluating the principles of imaging sensors which are exploited in the manufacture of commercial television cameras. The second section identifies an area which should be substantially larger than it is in dealing with such an important topic as 'special vision sensors'. Much more is needed in this area to satisfy the thirst for sensor applications in robotic workcells.

Fibre-optics plays an important part in robot vision and this is covered by a selection of papers in the third section. In the same way, the importance of laser technology in the design of sensors is covered in the fourth section.

A topic which is sometimes (conveniently) forgotten is that of scene illumination, including structured light. It cannot be stressed too strongly how important it is to control the illumination of the work area through the sense of innovative lighting techniques and, where possible, the use of structured light in the form of a 'light stripe'; this is the subject of section five.

It is the intention of the editor and the publisher to provide an easily accessible treatment of this crucial technology in anticipation that effort will be focused on providing commercial products more readily acceptable in robotic applications.

Professor Alan Pugh
University of Hull, UK

1

Reviews

The two papers in this section set the scene for robot sensors in general and imaging sensors in particular.

ROBOT SENSORS – A PERSONAL VIEW

A. Pugh
University of Hull, UK

The current situation surrounding robot sensors is surveyed in the context of sensory requirements for robotics. Visual and tactile sensors are covered in some detail with the deficiencies in existing sensory methods highlighted. The paper justifies the claim that there are very few sensors available which are designed specifically for robotic applications. In particular, it is often assumed that vision sensors are freely available, but the opposite is indeed the case.

What is the present situation in robotics? We see considerable maturity in the design and manufacture of the robot manipulator and in particular the evolution of new kinematic arrangements. Further, we see the acceptance of computer control systems to drive manipulators with a good initial attempt to introduce robot programming languages incorporating important interpolative functions. Consequently, it can be suggested that the mechanical manipulator coupled with its computer controller has reached a point of substantial development with virtual complete acceptance in a shop-floor environment.

Regrettably, environmental sensing is a different story. In parallel with the development of robotic technology, and possibly preceding it, we have a great understanding of computer vision invariably with applications in picture processing of one kind or another. It is assumed that there is natural integration of what has been researched and developed in computer vision with robotic technology but the marriage is an uneasy one. Indeed, a new industry is emerging from the specialised knowledge needed for the production of reliable sensory systems. Those involved in the application of environmental sensing to robotic installations rapidly understand the particular and peculiar requirements of the technology. Implementing real-time sensory feedback proves to be a difficult and challenging task requiring new thinking in the design of algorithms. And if this is not enough, the system builder does not have at his fingertips the availability of sensors which robotics demands despite the naive assumption that sensors – particularly vision sensors (the subject of this Volume) – are plentiful. It is against this background that the sensor crisis is explored here with an attempt to summarise the present position in terms of research and manufacture.

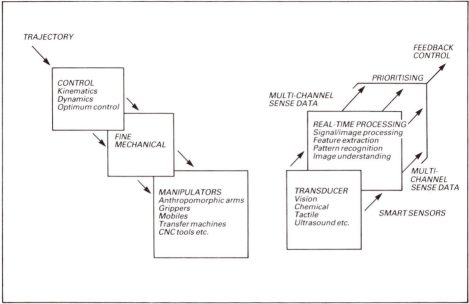

Fig. 1   The role of the manipulator and the sensor in robotics

To put the subject of robotics into context, Fig. 1 shows how robotics is seen through the eyes of the manipulator and alternatively through the eyes of the sensor. These diagrams give a clear pictorial description of these two aspects[1]. Davey completes the picture with a conceptual illustration of how the programmer might see the subject (Fig. 2) where the role of the sensor is clearly evident. This further underlines the crucial position occupied by the environmental sensor without which the total technology cannot be sustained.

Transducers available

The situation for vision transducers develops logically starting with the photodetector to determine presence (as realised in the proximity sensor) and progressively expanding the information collected through the linear array, which can be used either for range or one-dimensional transduction to the area array and television camera. These latter devices are perhaps most readily recognised as the fundamental transducers in vision systems. However, as we shall see later, there is more to this than meets the eye. The list for vision is completed by the recognition of the laser as an important tool in gathering three-dimensional information about the 'real world'. Whilst optical fibre is in itself not a transducer, it is a very powerful and important medium to communicate optical information away from a small volume.

The situation with tactile and force sensing is more confused. What is implied in force sensing is a measurement of pressure in the form of a moment about an axis or alternatively torsion concentric with an axis. When realised as a sensory wrist, the role of force sensing is readily apparent. Tactile sensing is better associated with tactile imaging when an attempt to replicate 'feel' through the construction of an array of tactile 'pressure

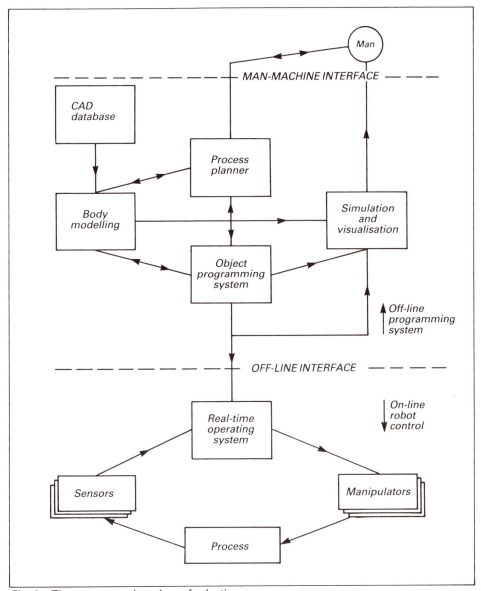

Fig. 2   The programming view of robotics

points' is used to construct shape information through touch. This latter class of sensors has resulted in some very interesting work but has barely reached commercial exploitation.

Acoustic transducers have a useful although specialised role to play in robotics. The promise of acoustic imaging is attractive – particularly for free-roving robots. This is a singularly difficult area in which to practice research and results are only now beginning to emerge from early experiments in this area. Apart from the use of microphones for voice control of robotic systems, they can for example be used to listen to characteristic noises emitted during the act of assembly[2].

Of course, a variety of other sensors might find application in robotics. Already, there is an example of the use of chemical sensing[3] although in this illustration the information extracted is not being used to guide the robot. The use of thermal imaging might have attractions if components can be heated or cooled relative to the ambient temperature and so enhance their 'visual' contrast. In the same way, ionising radiation might be used to enhance component shape. Indeed, the whole problem of scene illumination will be explored later.

Having taken a first look at the transduction principle available for sensors each sector will now be explored in turn in greater detail.

Vision sensors

Some misconceptions

It is generally believed that vision sensors are readily available to support robotics. The truth of the matter is that vision sensors are often confused with electronic cameras designed primarily for the television industry. Indeed, it must be stated, to avoid any misunderstanding, that there are virtually no vision sensors specifically manufactured for robotic applications! Having realised this fact, the robotic industry is in a dilemma in having the transducers available (through self-scanned imaging arrays) but having no cameras specifically designed for the robotic environment.

Again, the location of a vision sensor in the work area is frequently poorly implemented through a combination of the lack of suitable vision sensors coupled with an attempt to replicate the distribution of vision sensing in much the same way as the human form. All too often robot vision is incorporated by placing a solid-state camera above the work area just as our own eyes look down on the task implemented by our hands. It is of fundamental importance to understand that sensors must be distributed around the work area including the robot manipulator. By placing of a vision sensor on the end-effector of the robot, the problems of: parallax, coordinate transformation and resolution are greatly simplified.

Let us explore each of these in turn. A three-dimensional object when viewed as a two-dimensional profile will reveal a variation in perimeter size according to the angle of view[4]. The true profile shape of such a three-dimensional object can only be determined when the camera is placed immediately above its centre. With a camera placed on the end-effector, both the gripper and the camera are servoed to the centre line of the object and any parallax error first determined is eliminated through an iterative process as the end-effector approaches the centre-line.

Coordinate transformation between vision and robot axes is required whenever vision sensors are incorporated. Should the two be separated in space, the problem of coordinate transformation is considerable and the computational implications of the transformation matrix are well under-stood. However, all this presupposes that the calibration of both the manipulator and vision sensor can be maintained and sustained over a period of time. In a working environment this cannot be assumed. The expedient of placing the vision sensor on the end-effector reduces the problem of coordinate transformation to triviality and the physical separation of the mechanical gripper and the environmental sensor is determined by the dimension of the fixing bracket. Calibration is also far less critical with this form of implementation.

The popular belief that resolution should be as high as possible can only be sustained when the vision sensor is mounted remotely. It is very revealing to see how a low resolution sensor mounted on the end-effector can out-perform a much higher resolution camera mounted above the work area[5]. One of the earliest attempts to integrate properly the vision sensor with the robot gripper was published by the University of Nottingham[6]. The significance of this kind of thinking is not always understood – even after more than a decade of research and development.

Rebuilt cameras

The most elegant attempts to incorporate vision sensing in a robotic system are achieved by integrating a purpose-built camera with the robot end-effector. The best recent example arises from work at Oxford University[7] (see page 255). Such an approach is often implemented by taking an already manufactured solid-state camera and dismantling it to a point when it can be repackaged in a more robust structure. The Oxford welding head (manufactured by Meta Machines, UK) is enhanced by the knowledge that the working environment of the CCD array camera is dominated by a welding arc a mere 10cm away. Incidentally, this product also contains a laser light-stripe source of illumination which is discussed in the next section.

Another example of a repackaged camera has been published by the RCA Laboratories[8] (see page 85) on the design of an end-effector to handle loudspeaker assemblies. Incidentally, this particular end-effector also incorporates force feedback measured by the deflection of compliant beams. In looking at these two examples it would be natural to conclude that scope exists for a 'dismantled' camera system to be marketed for housing within custom-built end-effectors. This has an exact parallel in the availability of frameless dc motors commonly used as actuators in robot manipulators.

Some experiments on special vision sensors

Whilst special robotic vision sensors have not been made commercially in any quantity, a number of interesting experiments – sometimes conceptual – have been published. For example the very simple arrangement of a discrete array of parallel light beams shining across the space between gripper fingers has been investigated by the Bell Laboratories[9] (see page 139). When this simple arrangement of a relatively small number of light beams is integrated with the manipulating capability of the robot end-effector, useful information can be acquired about the components being handled. It should not be assumed that the relatively limited resolution implied by a few light beams dominates the resolving power of the total system. When combined with the fine positioning capability of a given manipulator, it is this resolution which also has an effect on the overall accuracy of the vision sensor. This is an extension of the optical proximity sensor and illustrates beautifully the concept of simple sensors used intelligently to present positional data to the robot controller.

Work at the Gould Research Centre[10] (see page 115) has turned upside down the concept of structured light. In this work optical fibres aligned in a coherent way are shaped geometrically at the imaging surface. This, then, imposes structure on the perceived image which itself can be used to enhance the feature information being extracted. Another relevant point to consider whilst evaluating coherent optical fibre bundles, is the need for small image sensors if integration with the end-effector is the objective. If it is not practical to include a camera at the end-effector, for one reason or the other, a coherent fibre optic bundle can be used to communicate the image away from the gripper. This. however, raises problems associated with mechanical robustness and cost of a coherent bundle.

In considering small cameras, it is timely to mention the small and rugged cameras used at the University of Hull to achieve the fundamental objective of close integration with the end-effector[11] (see page 67). The transducer used in these experiments is a dynamic RAM which has been encapsulated using a transparent cover so that it can be used as an imaging device. (In fact these are commercially available.) Perhaps the smallest camera using specially repackaged devices has been produced by Loughlin with the encapsulated sensor, complete with lens, being approximately the size of a thimble[12] (see page 95).

Work at Honeywell has exploited the range-finding properties of devices manufactured for domestic cameras to combine range-finding with area imaging in a single package. The image acquired by a single lens is divided so that a fraction is transmitted to an imaging CCD array and the remainder focussed onto the range-finding array[13] (see page 75). Experiments of this kind indicate very clearly the requirements for the robotic industry in terms of vision sensors. It is unfortunate that the market volume of the robotics industry is not sufficiently large to compete with other markets (notably television). Ultimately, the need for purpose-built vision sensors must result in commercially available products in significant numbers.

Scene illumination

The problem of contrast

Scene illumination is a neglected aspect of image recovery. The need for controlled illumination is well understood but so often the problem of feature extraction from perceived images is unnecessarily complicated by the lack of contrast in the image.

Backlighting has often been used to enhance the contrast to a point where segmentation through thresholding becomes possible. An elegant solution to this problem has been developed at the Philips Research Laboratories using parallel projection optics[14]. Incidentally this experiment also takes care of the problem of parallax by virtue of parallel projection in the work area. It will be noted here that the temptation to use a backlit table cannot be regarded as a universally accepted solution. An expedient such as that used at Nottingham in conjunction with coherent fibre-optic coupling is a much better implementation of this goal[15] (see page 147). Wherever it is possible, the vision sensor should look into a source of illumination of sufficient magnitude to swamp any ambient lighting.

Structured light

One of the most elegant solutions to the problem of scene illumination has been developed by General Motors[16] (see page 213). The novelty of the approach is well known and reported in the Consight system. In fact this development does tackle the problem of contrast although it appears as structured light.

Almost without exception, the exploitation of structured light is realised as a single light stripe projected onto the surface of the work area. The way that light structure behaves is well published[17,18] (see page 229). Experiments at SRI International illustrate the power of structured light in identifying overlapping three-dimensional components[19] (see page 245). A particularly good example of the use of structured light is that described previously where a light stripe is used to determine position of the weld path in the seam to be joined. The Meta Machines implementation projects a structured light stripe immediately ahead of the weld pool and the resultant image is perceived by the integrated camera.

Laser scanning

The concept of illuminating the work area using a laser beam coupled with a mechanical scanning arrangement provides the best way to recover three-dimensional information about the real world. It must be assumed that at this point in evolution such a laser scanning system cannot be incorporated within the end-effector of a robot. Consequently, it is common practice to associate laser scanning systems as free-standing sensors mounted above the work area. Used in this way, and given the limitations of mechanical scanning, the information content is very high indeed. If this information is used in conjunction with end-effector mounted sensors, the promise for a complete sensory understanding of the environment is probably the best available. Further, the problem of scene illumination is automatically solved when using a high intensity laser beam. A particularly good example of laser scanning by NRC in Ottawa can be found elsewhere[20] (see also page 175).

Force sensing

A problem removed

Some fundamental thinking at the Charles Stark Draper Laboratories[21] (see Vol. 2) provided an elegant and unexpected solution to the problem of correction of alignment during an insert operation. It was discovered by these laboratories that by careful kinematic design and the inclusion of adequate compliance, it was possible to provide a solution to this problem without the need for active servo-feedback. The device invented is well known as the remote-centre compliance and is available commercially from several companies. Further, the force sensing implied by this invention is now understood as passive sensing.

A development of the remote-centre compliance (RCC) has been presented by Carnegie–Mellon University which utilises the variable compliance attainable from a pressurised compliant system to control the parameters of the RCC device[22] (see Vol. 2). It might be said that the only commercially available sensors specifically designed for the robotic industry are realised in the RCC and compliant wrists fitted with electrical transducers.

Active force feedback

One of the earliest experiments in the use of a compliant wrist to provide error information for parts mating was presented by Hitachi[23] (see Vol. 2). The experimental prototype known as Hi-T-Hand Expert-1 was universally acclaimed as a significant demonstration of active compliant insertion. Similar experiments have been implemented in Leuven University in Belgium[24] (see Vol. 2) where a compliant wrist fitted with a six-component force sensor provides error information to a compatible multi-axis actuator.

This idea introduces the concept of micro-manipulation by attaching to the end of a relatively coarse robotic manipulator a device capable of finer precision. This principle is inverted in an experiment conducted by Hitachi[25] (see Vol. 2) where the concept of a sensory table mounted beneath the robot and equipped with a six-axis force sensor is used to compensate for misalignment during the parts-mating process.

Tactile imaging

A first look at the problem

The objective is the recognition of shape through touch with the additional possibility of sensing pressure applied during the act of gripping. The concept is simple to understand – an array of pressure sensors integrated beneath a skin surface with sufficient size and resolution to be compatible with the task to be accomplished. However, experiments published to date have been distinctly dominated by the limitations in the technology used to fabricate the sensory pad. In looking at the fabrication of tactile sensors, the immediate problem is that of selection of a suitable material to perform the transduction process. Carbon-loaded rubbers and polymers have featured prominently in various experiments although it will be discovered that the response or settling time coupled with the recovery time and hysteresis are impossibly long for realistic application. This problem is not always recognised in prototype devices which have been built. There is, however, evidence that a breakthrough in approach is imminent with the understanding that the reality of a reliable commercial device is still a long way ahead.

Orthogonal arrays

An experiment by Purbrick[26] is impressive in its simplicity. An orthogonal array of round carbon-loaded rubber strips form resistive contacts at the intersections which change resistance with pressure at each tactile site (taxel). A similar experiment conducted at the Hirst Research Centre in the UK[27] (see Vol. 2) uses an orthogonal array of electrodes sandwiching a resistive mat comprising carbon-loaded cotton. Results published on the sensor show the reliability in performance expected from a practical device and the problem of current spreading in surrounding tactile sites is addressed using feedback techniques.

Discrete arrays

In one form or another these sensors are fabricated from discrete probes assembled sufficiently close together to provide the function of a tactile skin. The Lord Corporation[28] (see Vol. 2) has fabricated such an array using the interruption of transmitted light at each tactile site by a small probe pressed into the structure of the sensor. This approach is fundamentally rugged but inherently limited in resolution. An earlier experiment at the University of Nottingham, UK[29] (see Vol. 2) demonstrated an 8 × 8 array of probes which could be pushed back into the body of the sensor by 2.5cm. A sensing plane determined the first increment of motion of each probe and the probe position was updated subsequently by knowledge of robot end-effector motion. This sensor had the capability of feeling three-dimensional objects within the size of $2.5cm^3$.

A conceptual idea reported by the Bell Laboratories[30] (see Vol. 2), shows a 7 × 7 array using magnetoresistive transduction at each tactile site. This contribution argues that the mere determination of shape through touch does not represent the fundamental goal of such a sensor. The authors argue that this information can be achieved in other ways and that a tactile array sensor should concentrate on the determination of forces such as shear and torque. It is argued that no fewer than five discrete tactile measurements are needed to be transduced at each tactile site. Clearly, this approach represents more accurately the information which can be transduced between human fingers and would provide very valuable sensory data to any robotic controller. At present, robotic system designers would be grateful for any form of tactile transduction no matter how primitive.

Integrated arrays

A singularly impressive experiment has been reported from Carnegie-Mellon University[31] (see Vol. 2) where an integrated array of processors on a silicon substrate separated by active and passive metallic electrodes form the tactile sensor in conjunction with a carbon-loaded skin placed over the surface. The design of the processor array provides communication of sensory information between adjacent tactile sites. This, together with instructions imposed on the sensor, enables features to be extracted directly through the principle of parallel array computation. This is a true example of the implementation of a smart sensor.

An alternative use of silicon has recently been reported[32] (see Vol. 2) where a tactile site is etched out of a silicon substrate by forming a compliant silicon beam on which are integrated strain measuring elements. This technique will provide an integrated array of tactile beams of sufficient resolution to provide a useful and rugged tactile sensor. This use of a silicon substrate to combine the function of a compliant beam with monolithic integrated electronics is indeed intriguing and one which offers considerable promise.

Some recent tactile experiments

Piezoelectric polymers, in particular PVF_2, are finding application in tactile sensors by measuring the change in the acoustic propogation constant of a compliant material with pressure. The property of PVF_2 has been reported[33] and it is believed that some experimental sensors based on this principle have been constructed.

Two research groups have discovered the effect of the scattering of light from an acrylic plate when contact is made with a membrane material. This technique has the promise of very fine resolution indeed, limited only by the resolving power of the vision sensor needed to transduce the reconstructed tactile information through the optical medium. The sensor comprises a tactile membrane separated by a small distance from an acrylic plate into which light is injected. When the membrane touches the acrylic plate in response to a tactile pressure image, light is scattered from the acrylic plate in proportion to the pressure applied at the point of contact. Consequently, a high-resolution optical image can be seen representing directly the tactile image of a given object[34, 35] (see Vol. 2). One fascinating feature of this kind of sensor is that for an optically transparent membrane, the imaging array can be used for both tactile and visual sensing of the same object.

Concluding remarks

This paper is an attempt to outline the present position in sensors for robotic applications. It cannot be emphasised too strongly that there is a chronic shortage of sensors commercially available which are truly suitable for robotic applications. Further, the raw information which is extracted by the sensory transducer requires sensible manipulation before useful geometric data are available. A brief reference has been made to smart sensors. The combined objective of integrating the sensory transducer with suffcient computational support directly to implement feature extraction imposes an artificial difficulty which is really unnecessary. Researchers are naturally attracted by the intellectual challenge of smart sensors although the same goal is achieved even if the transducer and processor are separated in distance. Smart sensing must be interpreted liberally and as much effort expended on successful transduction as on computation support for fast real-time feature extraction.

There is no doubt at all that despite the very large investment in research effort on the subject of robotics over the past five years, the technological challenge imposed by environmental sensors is proving to be many orders of magnitude more difficult than might have been supposed a decade ago. It is indeed encouraging to find a handful of new transduction ideas emerging – particularly in the area of tactile imaging. If we are to enter the third generation of robotics, the availability of cheap, reliable and rugged sensors is absolutely essential. Without these, the whole technology founders. Industrial applications of visual and tactile sensing are really in the experimental stage and it is salutory to confess that the only universally accepted and reliable sensor which industry can use is the proximity sensor.

Acknowledgement

The author acknowledges with gratitude permission to reproduce Figs. 1 and 2 from material previously written by Mr P. G. Davey of Meta Machines Limited and Oxford University in the UK.

References

[1] Davey, P. G. 1983. Robots with commonsense? The research scene today. *Journal of the Royal Society of Arts*, October: 671-685.

[2] Smith, R. C. and Nitzan, D. 1983. A modular programmable assembly station. In, *Proc. 13th Int. Symp. on Industrial Robots*, pp. 5.53-5.75. SME, Dearborn, MI, USA.

[3] Robots to sniff out those water leaks. *The Industrial Robot*, 9(3): 150-152, 1982.

[4] Carlisle, B., Roth, S., Gleason, J. and McGhie, D. 1981. The Puma/VS – 100 Robot Vision System. In, *Proc. 1st Int. Conf. on Robot Vision and Sensory Controls*, pp. 149-160. IFS (Publications) Ltd, Bedford, UK.

[5] Hill, J. J., Burgess, D. C. and Pugh, A. 1984. The vision-guided assembly of high-power semiconductor diodes. In, *Proc. 14th Int. Symp. on Industrial Robots*, pp. 449-460. IFS (Publications) Ltd, Bedford, UK.

[6] Heginbotham, W. B., Gatehouse, D. W., Pugh, A., Kitchin, P. W. and Page, C. J. 1973. The Nottingham SIRCH assembly robot. In, *Proc. 1st Conf. on Industrial Robot Technology*, pp. 129-138. IFS (Publications) Ltd, Bedford, UK.

[7] Morgan, C. G., Bromley, S. J. E., Davey, P. G. and Vidler, A. R. 1983. Visual guidance techniques for robot arc-welding. In, *Proc. 3rd Int. Conf. on Robot Vision and Sensory Controls*, pp. 615-633. IFS (Publications) Ltd, Bedford, UK.

[8] Baird, H. S. and Lurie, M. 1983. Precise robot assembly using vision in the hand. In, *Proc. 3rd Int. Conf. on Robot Vision and Sensory Controls*, pp. 533-539. IFS (Publications) Ltd, Bedford, UK.

[9] Beni, G., Hackwood, S. and Rin, L. 1983. Dynamic sensing for robots – An analysis and implementation. In, *Proc. 3rd Int. Conf. on Robot Vision and Sensory Controls*, pp. 249-255. IFS (Publications) Ltd, Bedford, UK.

[10] Agrawal, A. and Epstein, M. 1983. Robot eye-in-hand using fiber optics. In, *Proc. 3rd Int. Conf. on Robot Vision and Sensory Controls*, pp. 257-262. IFS (Publications) Ltd, Bedford, UK.

[11] Whitehead, D. G., Mitchell, I. and Mellor, P. V. 1984. A low-resolution vision sensor. *J. Phys. E: Sci. Instrum.*, 17: 653-656.

[12] Loughlin, C., Morris, J., Rovetta, A. and Franchetti, I. 1984. Line, edge and contour following with eye-in-hand vision system. In, *Proc. 14th Int. Symp. on Industrial Robots*, pp. 553-559. IFS (Publications) Ltd, Bedford, UK.

[13] Orrock, J. E., Garfunkel, J. H. and Owen, B. A. 1983. An integrated vision range sensor. In, *Proc. 3rd Int. Conf. on Robot Vision and Sensory Controls*, pp. 263-269. IFS (Publications) Ltd, Bedford, UK.

[14] Saraga, P. and Jones, B. M. 1981. Parallel projection optics in simple assembly. In, *Proc. 1st Int. Conf. on Robot Vision and Sensory Controls*, pp. 99-111. IFS (Publications) Ltd, Bedford, UK.

[15] Cronshaw, A. J., Heginbotham, W. B. and Pugh, A. 1979. Software techniques for an optically-tooled bowl-feeder. In, *Proc. IEE Conf. on Trends in On-Line Computer Control Systems*, Vol. 172, pp. 145-150. IEE, London.

[16] Holland, S. W., Rossol, L. and Ward, M. R. 1979. Consight-1: A vision-controlled robot system for transferring parts from belt conveyors. In, *Computer Vision and Sensor-Based Robots* (Eds. Dodd, G. G. and Rossol, L.), pp. 81-100. Plenum Press, New York.

[17] VanderBrug, G. J., Albus, J. S. and Barkmeyer, E. 1979. A vision system for real-time control of robots. In, *Proc. 9th Int. Symp. on Industrial Robots*, pp. 213-231. SME, Dearborn, MI, USA.

[18] Schroeder, H. E. 1984. Practical illumination concept and technique for machine vision applications. In, *Robots 8 Conf. Proc*, pp. 14.27-14.43. SME, Dearborn, MI, USA.

[19] Bolles, R. C. 1981. Three-dimensional locating of industrial parts. In, *8th NSF Grantees Conf. on Production Research and Technology*, pp. W1-W5 National Science Foundation, Washington, DC, USA.

[20] Rioux, M. 1984. Compact, highspeed 3-D vision sensor for robotic applications. NRC, Ottawa, Ontario, Internal paper.

[21] Whitney, D. E. and Nevins, J. L. 1979. What is the remote center compliance (RCC) and what can it do?. In, *Proc. 9th Inst. Symp. on Industrial Robots*, pp. 135-152. SME, Dearborn, MI, USA.

[22] Cutkosky, M. R. and Wright, P. K. 1982. Position sensing wrist for industrial manipulators. In, *Proc. 12th Int. Symp. on Industrial Robots*, pp. 427-438. IFS (Publications) Ltd, Bedford, UK.

[23] Goto, T., Inoyama, T. and Takeyasu, K. 1974. Precise insert operation by tactile controlled robot. In, *Proc. Conf. on Industrial Robot Technology*, pp. C1.1-C1.8 IFS (Publications) Ltd, Bedford, UK.

[24] Van Brussel, H. and Simons, J. 1978. Automatic assembly by active force feedback accommodation. In, *Proc. 8th Int. Symp. on Industrial Robots*, pp. 181-193. IFS (Publications) Ltd, Bedford, UK.

[25] Kasai, M. et al. 1981. Trainable assembly system with an active sensory table possessing 6 axes. In, *Proc. 11th Int. Symp. on Industrial Robots*, pp. 393-404. JIRA, Tokyo.

[26] Purbrick, J. A. 1981. A force transducer employing conductive silicone rubber. In, *Proc. 1st Int. Conf. on Robot Vision and Sensory Controls*, pp. 73-80. IFS (Publications) Ltd, Bedford, UK.

[27] Robertson, B. E. and Walkden, A. J. 1983. Tactile sensor system for robotics. In, *Proc. 3rd Int. Conf. on Robot Vision and Sensory Controls*, pp. 327-332. IFS (Publications) Ltd, Bedford, UK.

[28] Rebman, J. and Morris, K. A. 1983. A tactile sensor with electro-optical transduction. In, *Proc. 3rd Int. Conf. on Robot Vision and Sensory Controls*, pp. 341-347. IFS (Publications) Ltd, Bedford, UK.

[29] Sato, N., Heginbotham, W. B. and Pugh, A. 1977. A method for three dimensional part identification by tactile transducer. In, *Proc. 7th Int. Symp. on Industrial Robots*, pp. 577-585. JIRA, Tokyo.

[30] Hackwood, S., Beni, G. and Nelson, T. J. 1983. Torque-sensitive tactile array for robotics. In, *Proc. 3rd Int. Conf. on Robot Vision and Sensory Controls*, pp. 363-369. IFS (Publications) Ltd, Bedford, UK.

[31] Raibert, M. H. and Tanner, J. E. 1982. A VLSI tactile array sensor. In, *Proc. 12th Int. Symp. on Industrial Robots*, pp. 417-425. IFS (Publications) Ltd, Bedford, UK.

[32] Allan, R. 1984. Sensors in silicon. *High Technology*, September: 43-50.

[33] Dario, P. et al. 1983. Piezoelectric polymers: New sensor materials for robotic applications. In, *Proc. 13th Int. Symp. on Industrial Robots*, pp. 14.34-14.49. SME, Dearborn, MI, USA.

[34] Mott, D. H., Lee, M. H. and Nicholls, H. R. 1984. An experimental very high resolution tactile sensor array. In, *Proc. 4th Int. Conf. on Robot Vision and Sensory Controls*, pp. 241-250. IFS (Publications) Ltd, Bedford, UK.

[35] Tanie, K. et al. 1984. A high resolution tactile sensor. In, *Proc. 4th Int. Conf. on Robot Vision and Sensory Controls*, pp. 251-260. IFS (Publications) Ltd, Bedford, UK.

ROBOT VISION: AN EVALUATION OF IMAGING SENSORS

L. J. Pinson
University of Colorado, USA

Robot vision, robotics vision, and computer vision are terms that have evolved over the past few years to include two separate but related functions that provide visual sensing for computer-driven robots. These two functions are 'electro-optical imaging' and 'image processing'. The purpose of electro-optical imaging is to convert optical radiation to an appropriate electronic signal for input to the robot's computer; whereas, the purpose of image processing is to extract useful information from the electronic image provided by the sensor. This article deals with the electro-optical sensing part of robot vision. It describes the operation and properties of electro-optical imaging sensors. The fundamental concepts of optical radiation, optical radiation quantities and units, and photon energies are defined. The fundamental principles for detecting optical radiation and definitions for the primary performance measures for optical detectors are given. A relatively comprehensive overview of the parameters used to describe imaging sensors is presented. Factors affecting these parameters and obtainable sensor performance are compared. Example design analyses are presented to show the interaction of various performance measures.

Robot vision, robotics vision, and computer vision are terms that have evolved over the past few years to include two separate but related functions that provide visual sensing for computer-driven robots. These two functions are 'electro-optical imaging' and 'image processing'. The purpose of electro-optical imaging is to convert optical radiation to an appropriate electronic signal for input to the robot's computer; whereas, the purpose of image processing is to extract useful information from the electronic image provided by the sensor. The image processing function is perhaps more difficult because it must answer the questions: (1) What constitutes useful information? and (2) What algorithms will efficiently extract that information from the image?

This article deals with the electro-optical sensing part of robot vision. It describes the operation and properties of electro-optical imaging sensors.

The requirements for an imaging sensor are dependent upon the sophistica-tion of image processing methods. There will be a demand for imaging sensors that provide higher resolution, more pixels, and multispectral imagery. All these demands result in a larger amount of information to be handled by the processor. A relatively comprehensive overview of the parameters used to describe imaging sensors is presented. Factors affecting these parameters and obtainable sensor performance are compared.

Fundamentals

Robot vision, or electro-optical imaging, can be very simple or very complex; but, in all cases, it involves the 'detection' of optical energy. Detection in most cases means the conversion of optical energy to an electrically measurable parameter such as current, voltage, resistance, capacitance, or charge. This fundamental concept is illustrated schematic-ally in Fig. 1. The degree of complexity for any given robot vision system is driven by objectives for the total robot system. Objectives may be as simple as the location of a few reflectors or light sources on an object or as complex as collecting real-time, high-resolution, multispectral imagery. Furth-ermore, the operational environment may be controlled (as with identifying parts on a conveyor belt) or unknown (as with outdoor terrain). In general, many factors affect the complexity and 'intelligence' required of a robot system and these will directly impact the sophistication required of the imaging sensor.

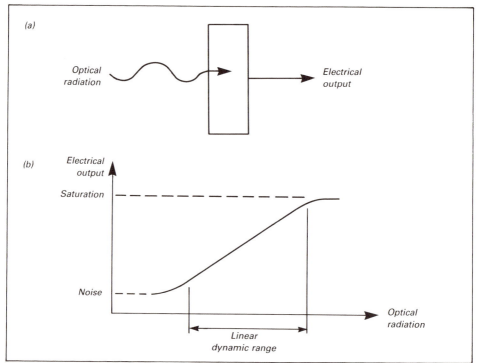

Fig. 1   *Detection of optical radiation: (a) detection and (b) dynamic range*

The optical spectrum

The input to an imaging sensor is an optical field characterised as having properties of the 'optical spectrum'. The optical spectrum is defined to be a subset of the electromagnetic spectrum that includes 'optical' wavelengths. The range of wavelengths included in the optical spectrum is not universally agreed upon nor are there rules for determining fixed boundaries within the electromagnetic spectrum. The optical spectrum is defined as including wavelengths from 10nm to 1mm (or 0.01–1000μm)[1]. Of this wavelength range the predominant usage in electro-optical systems is in the subrange from about 0.1 to 50μm.

Fig. 2 shows specific regions of the electromagnetic spectrum and identifies various subregions of the optical spectrum. The visible portion of the optical spectrum is identified as the range of wavelengths from approximately 0.380 to 0.760μm. Other labelled subregions in the optical spectrum such as ultraviolet (uv), near, middle, and far infrafed have somewhat arbitrary boundaries that have evolved from practical experience and specific applications areas over a number of years.

Optical radiation coming from an object has two possible origins. One is from internal activity of the atoms composing the object. Energies corresponding to wavelengths in the optical spectrum typically involve transitions of electrons within the atom. As the electrons change energy levels they absorb and/or emit quanta of energy (photons) in the form of

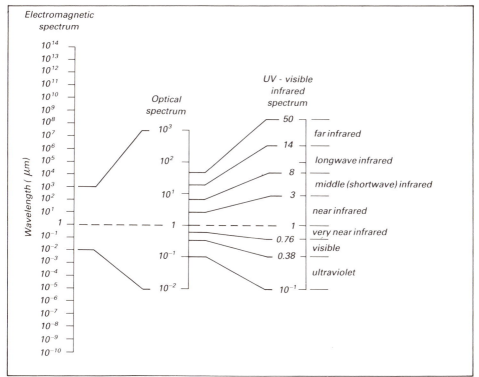

Fig. 2   The optical spectrum

electromagnetic radiation. The electrons can be stimulated to make transitions by internal energy, by chemical reactions, or by external sources of energy (e.g. electromagnetic fields). The intensity and wavelengths of the emitted radiation depend on the nature of the stimulation.

A second source of radiation from an object is reflection or transmission of other radiant sources in the ambient environment of the object. Reflection can be due to simple scattering or it may involve absorption and re-emission of selected wavelengths. A transmitted wavelength is one that passes through the object and is not considered to be a part of the object's optical field.

Radiometric, photometric, and quantum units

Radiometric quantities are defined in terms of the mks system of units. Since photons have an energy given by hc/λ, the spectral radiometric quantities may be expressed in terms of either energy or photon fluxes (photons/second). This alternative set of radiometric quantities is useful in evaluating electro-optical systems where quantum effects are important.

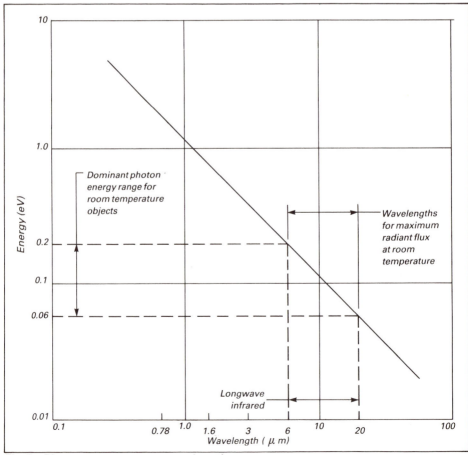

Fig. 3   Photon energy versus wavelength

For example, several noise processes in quantum detectors are described in terms of photon flux. A common practice to distinguish photon flux representations of radiometric quantities is to use a subscript e with each of the radiometric symbols. This is a little cumbersome, however, since spectral quantities (wavelength dependent) also use a subscript λ. For reasons of simplicity the e subscript is not used. The same symbols are used for both energy units and photon flux units. Actual units will either be clear from the development or will be stated.

Another set of units that is commonly used to describe energy levels in semiconductors is based on electron volts (eV), where 1 eV $= 1.602 \times 10^{-19}$ J. Thus, the energy of a photon can now be expressed either in joules or electron volts. For λ in micrometres:

$$Q_{photon} = \frac{hc}{\lambda} \text{ (J)} = \frac{hc}{\lambda(1.602 \times 10^{-19})} \approx \frac{1.25}{\lambda} \text{ (eV)} \qquad (1)$$

Fig. 3 shows photon energy in electron volts versus wavelength. This figure is useful in understanding how the response of semiconductor detectors to optical radiation is dependent on wavelength, as related to the transition energy gaps for charge carriers to reach the conduction band. A photon colliding with a charge carrier must impart enough energy to cause that charge carrier to jump to the conduction band.

It has been stated that photometric units are based on the eye response of an 'average young human'. Furthermore, photometric units were developed to quantify visible light. Two spectral response curves for the human eye are given that represent relative response versus wavelength under strong (photopic) and reduced (scotopic) levels of radiant energy. Photopic response is typical of daylight vision and scotopic response is typical of night vision. These relative spectral response curves are shown in Fig. 4a. It is clear from the figure that the human eye response is more sensitive and shifts to shorter wavelengths at the reduced level of radiant energy. The photopic response represents the response of cones in the eye, which are predominantly distributed in the foveal or central part of the eye. The scotopic response is for the rods, which are distributed primarily in the peripheral part of the eye. Colour vision is a function of the cones so that the increased sensitivity of night vision lacks the ability to discriminate colours well.

The curves shown in Fig 4b are plots of spectral luminous efficacy for photopic and scotopic vision with units of lumens per watt (lm/W). These curves may be used to convert radiometric units to equivalent photometric units. If a subscript v is used to denote visible or photometric quantities then the photometric flux in lumens from a source is given by:

$$\phi_v = \int_0^\infty K(\lambda) \cdot \phi(\lambda) \, d\lambda \text{ (lm)} \qquad (2)$$

Additional details on the conversion between photometric and radiometric units are given in the Appendix along with an explanation of the maximum values of spectral luminous efficacy shown in Fig. 4b.

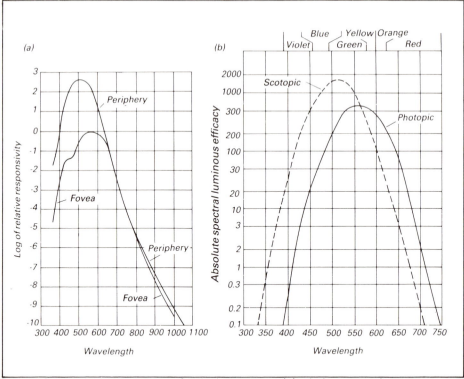

Fig. 4   Spectral response curves for the human eye: (a) relative spectral response (courtesy of Dr. G. Wald and the Optical Society of America), (b) spectral luminous efficacy (courtesy of RCA Corporation)

Table 1 lists corresponding radiometric and photometric quantities as well as symbols and names used for radiometric quantities before SI standardisation.

Optical radiation signatures: definition and calculation

The concept of an optical radiation 'signature' is useful in describing the total radiant flux from an object or source that is part of an electro-optical system. This optical signature consists, in most cases, of two major parts. It includes radiant flux emitted by the object and radiant flux from external sources that is reflected by the object. Both the emitted flux and the reflected flux depend on a number of different parameters which must be known if an accurate signature is to be obtained.

Details of the emitted radiant flux depend on whether the source is passive (e.g. a black-body) or active (LED, lamp, laser). The reflected radiant flux depends on the spectral reflectance of the source/object and on what ambient sources irradiate the source/object (sun, lamp, laser). For active sources, the dominant component is expected to be the emitted flux. However, for the passive object/source, both emitted and reflected ambient flux can be important. Their relative importance depends on the wavelength region of interest, relative temperatures, and the presence of ambient sources. For example, an electro-optical system designed for daytime use in

Table 1 Optical radiation quantities and units

Quantity	Symbol	Definition	Units	Symbol
Radiometric quantities and units (SI)				
Radiant energy	Q	Basic unit	joule	J
Radiant flux	ϕ	$\phi = \delta Q/\delta t$	watt	W
Radiant density	w	$w = \delta Q/\delta V$	joule/volume	J/m^3
Radiant intensity	I	$I = \dfrac{\delta\phi}{\delta\omega} = \dfrac{\delta^2 Q}{\delta t\delta\omega}$	watts/steradian	W/sr
Radiant flux density at a surface				
Radiant exitance (emitted)	M	$M = \delta\phi/\delta A$	watts/area	W/m^2
Irradiance (incident)	E	$E = \delta\phi/\delta A$	watts/area	W/m^2
Radiance	L	$L = \delta^2\phi/\delta\omega\delta S$	watts/ster-area	W/sr m^2
Emissivity	ε	$\varepsilon = \dfrac{Q(\text{emitted})}{Q(\text{blackbody})}$	numeric	
Absorptance	α	$\alpha = \dfrac{Q(\text{absorbed})}{Q(\text{incident})}$	numeric	
Reflectance	ρ	$\rho = \dfrac{Q(\text{reflected})}{Q(\text{incident})}$	numeric	
Transmittance	τ	$\tau = \dfrac{Q(\text{transmitted})}{Q(\text{incident})}$	numeric	
Radiometric quantities and units (before SI)*				
Radiant energy	U	joule		J
Radiant power	P	watts		W
Radiant energy density	u	joule/volume		J/cm^3
Radiant intensity	J	watt/steradian		W/sr
Radiant emittance (emitted)	W	watt/area		W/cm^2
Irradiance (incident)	H	watts/area		W/cm^2
Radiance	H	watts/steradian area		W/sr cm^2
Photometric quantities and units				
Luminous energy	Q_v	$Q_v = \int K(\lambda)Q(\lambda)\,d\lambda$	lumen second	lm s
Luminous flux	ϕ_v	$\phi_v = \delta Q_v\delta t$	lumen	lm
Luminous density	w_v	$w_v = \delta Q_v/\delta V$	lumen second/vol	lm s/m^3
Luminous intensity	I_v	$I_v = \delta\phi_v/\delta\omega$	lumen/steradian (candela)	lm/sr
Luminous flux density at a surface				
Luminous exitance (luminous emittance)	M_v	$M_v = \delta\phi_v/\delta A$	lumen/area	lm/m^2
Illuminance (illumination)	E_v	$E_v = \delta\phi_v/\delta A$	lux (lm/m^2)	lx
Luminance (brightness)	L_v	$L_v = \delta^2\phi_v/\delta\omega\delta S$	nit (candela/m^2)	nt
Spectral luminous efficacy	$K(\lambda)$	$K(\lambda) = \phi_v(\lambda)/\phi(\lambda)$	lumen/watt	lm/W
Spectral luminous efficiency	$V(\lambda)$	$V(\lambda) = K(\lambda)/K_{max}(\lambda)$	numeric	

* Spectral quantities add subscript λ to the symbol and per micrometre to the units

the visible wavelength band (e.g. a TV camera) relies almost entirely on reflected sunlight for its optical signal. For indoor or night time use the ambient sunlight is replaced by high-intensity lamps; however, the optical signature is still reflected ambient flux.

Optical signatures for active sources are typically given in terms of spectral radiance or spectral radiant exitance curves provided by the manufacturer. These curves can be integrated over the wavelength region of interest to obtain radiant flux in watts or photons/second. These manufacturer-provided spectral curves are typically obtained from measurement data for a specific set of operating conditions.

Optical emitted signatures from passive sources are usually obtained either by measurement or by modelling the source as an approximate black-body. Reflected signatures from passive sources are based on spectral reflectance of the object as well as spectral radiant exitance and relative geometries of ambient sources. The amount of detail included in an approach of this type ranges from very simple to very complex, depending on the amount of resources available and the importance of precise answers. Complex models exist that use data on object materials, geometry, and multiple ambient sources.

Rigorous theoretical definitions are given below for optical signatures along with calculable results based on simplifying approximations. The optical signature (OS) for an object is defined as the sum of emitted and reflected radiant flux (in watts or photons/second):

$$OS = \phi_e + \phi_r \text{ (W or photons/s)} \tag{3}$$

Emitted radiation. A typical object/source will consist of a number of different materials with different areas, spectral properties, and temperatures. A number of complex objects of interest have been evaluated experimentally to measure their optical signatures in selected spectral regions. Certainly this is true for designed optical sources such as lasers, LEDs, and discharge lamps. Experimental signatures have been obtained for a wide variety of other sources as well, including rocket plumes, vehicles of various types, and terrain, to mention a few.

Consider an object where the elemental area dA in a given viewing direction has a constant temperature, a single-material composition, and fixed spectral properties. Then the total emitted radiant flux from the object is obtained by integrating over the surface area of the object:

$$\phi_e = \int_{\text{surface}} dA \int_0^\infty d\lambda \cdot M_\lambda (dA) \tag{4}$$

where $M_\lambda (dA)$ is the spectral radiant exitance from the subarea dA and is dependent on its temperature as well as its spectral emissivity. In general, an expression for M_λ that shows dependence on changing parameter values across the object surface has too many unknowns for practical computation. Given this limitation, ϕ_e can be approximated by using known parameters for the object and by assuming values for unknown parameters.

Further simplifications are possible for many applications because (1) there is interest in only a narrow range of wavelengths, and (2) many objects of interest have a surface composed of only one or a few materials. For a narrow range of wavelengths many materials behave essentially as grey-body radiators (emissivity is constant for all wavelengths). Based on the second statement above, it is possible to assume a uniform temperature across the object surface. If the dominant surface material is known, then a value for emissivity can be found or assumed. All these specifications lead to an analytically calculable emitted flux:

$$\phi_e = A\varepsilon_A \int_{\lambda 1}^{\lambda 2} M_\lambda(T_A)\, d\lambda \qquad (5)$$

where ε_A and T_A are the object emissivity and temperature, respectively, A is the total surface area, and the integral gives total radiant flux from a black-body at the given temperature in the wavelength band $\lambda_1 \rightarrow \lambda_2$.

Reflected ambient radiation. Detailed characterisation of reflected ambient radiation is even more complicated than the emitted radiance from an object/source. In addition to dependence on the object/source spectral properties, the reflected radiant flux depends on the number, spectral properties, and geometric orientation of ambient sources. The problem becomes tractable for most applications only if they deal with one or a few ambient sources.

Ambient sources include a number of natural environment sources (e.g. sun, moon, stars, earth) as well as active sources designed to be a part of the system (e.g. lasers, floodlights, etc.). Careful design of an electro-optical system for a particular application will enhance the effect of desired ambient sources and minimise the effect of undesired ambient sources. This careful design is accomplished by appropriate selection of wavelengths, by consideration of relative geometries among the system components and ambient sources, and by shielding and baffle designs that restrict unwanted sources of radiation from entering the system.

Consider the spectral irradiance $E_{\lambda,m}$ from the mth ambient source in a particular environment incident upon a complex object source. The reflected radiant flux is given by:

$$\phi_r = \sum_m \int dA \int_0^\infty d\lambda\, E_{\lambda,m}\,(dA)\, \rho_\lambda\,(dA) \qquad (6)$$

where it is implicit that the quantities $E_{\lambda,m}$ and ρ_λ depend also on relative geometries between the ambient source and the surface area dA.

Measurement data that give spectral radiant exitance for the primary ambient sources (sun, moon, earth, stars) are available in selected references[2,3]. These data are typically dependent on a number of measurement parameters. For example, solar spectral radiation at the earth's surface depends strongly on the spectral properties of the atmosphere since that radiation must pass through the atmosphere to reach the earth's surface. Spectral properties of the atmosphere depend on the constituent gases, the amount of water vapour, and the concentration of particulate matter (e.g. dust). These parameters show a wide range of variability so that many different curves are obtainable that show solar spectral radiation at the earth's surface.

By approximating ambient sources as grey-bodies, treating the optical path properties separately, and assuming the object reflectance is constant over a limited wavelength region, the reflected radiant flux is then calculable from Eqn. (6) as:

$$\phi_r = \sum_m A \cdot \rho_A \int_{\lambda_1}^{\lambda_2} g_m \cdot M_{\lambda,m} \cdot \tau_\lambda\,(\text{path}) \cdot d\lambda \qquad (7)$$

where ρ_A is the constant reflectance over the object area. Furthermore if the range of wavelengths is small, then the path transmittance, τ_λ (path) may be approximated as a constant τ_p and the problem of computing reflected radiant flux becomes simple computationally:

$$\phi_r = \sum_m g_m \cdot A \cdot \rho_A \cdot \tau_p \int_{\lambda_1}^{\lambda_2} M_{\lambda,m} \cdot d\lambda \qquad (8)$$

$$= \sum_m g_m \cdot M_{\lambda_1 - - \lambda_2, m} \cdot A \cdot \rho_A \cdot \tau_p$$

where g_m is a geometric factor that will be different for each ambient source. In the most general case g_m will depend on relative geometries of the ambient source and the object. Furthermore the reflectance will in general depend not only on wavelength but on relative geometries as well. For a specular surface, the reflectance is strongly dependent upon reflection angle. For a diffuse (or Lambertian) surface it can be shown that the radiance is independent of angle so that approximation in Eqn. (8) is valid only for diffusely reflecting surfaces.

Geometry effects and the optical path. The results given for emitted and reflected optical signatures in the previous two sections were in terms of radiant flux (W/m²) or (photons/s-m²). Furthermore, the reflected signature contained a geometric term g_m that translated radiant exitance from an ambient source into irradiance at the object.

Consider the situation shown in Fig. 5. The object has an optical signature consisting of both emitted and reflected flux. It is desired to

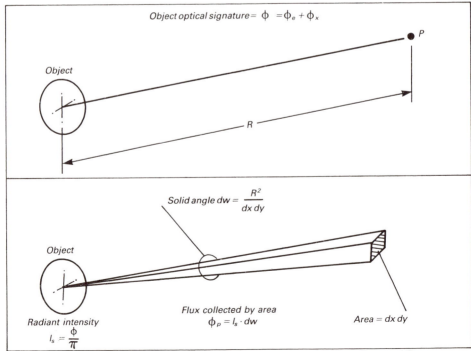

Fig. 5   *Geometry effects on radiant flux: (a) geometry and (b) irradiance at P*

measure the irradiance E_p, at observation point P a distance R from the object. If the object is Lambertian, then the radiant intensity in a given direction is:

$$I_\theta = I_n \cdot \cos\theta = \frac{\phi}{\pi}\cos\theta = \frac{M \cdot dA \cdot \cos\theta}{\pi} \tag{9}$$

The dependence on θ is most easily incorporated into the term for area; that is, the projected area of dA in viewing direction θ is given by $dS = dA \cos\theta$. Thus if the radiant flux from an object is computed from the object's cross-sectional area in a given direction, then the radiant intensity becomes independent of direction. For a Lambertian source the result is from Eqns. (5) and (8):

$$\phi_e = S\varepsilon_s M_{\lambda_1 - \lambda_2} \tag{10}$$

and

$$\phi_r = \sum_m g_m M_{\lambda_1 - \lambda_2, m} S\rho_s \tau_p \tag{11}$$

where the object area has been replaced by the cross-sectional area in a given direction, denoted by S.

Next, it is desired to find the irradiance of point P due to radiant flux from the object. First, compute ϕ_e and ϕ_r from Eqns. (10) and (11), then the irradiance by the object of an elemental area $dx\, dy$ at point P is given as:

$$E_p = \frac{\phi_p}{dx\, dy} = \frac{I_s\, d\omega}{dx\, dy}\,(\text{W/m}^2) \tag{12}$$

or

$$E_p = \frac{I_s}{R^2} = \frac{\phi}{\pi R^2}\,(\text{W/m}^2) \tag{13}$$

Eqn. (13) gives a simple relationship between radiant flux from a Lambertian source and irradiance at a distance R metres from the source. It should be clear that the geometeric factor g_m in the equations for ϕ_r is given by (assuming the ambient sources are Lambertian):

$$g_m = S_m / \pi R_m^2 \tag{14}$$

where S_m is the cross-sectional area of source m in the direction of the object and R_m is the distance between source m and the object.

Optical detectors

Most optical detectors for electro-optics applications can be placed in one of two general classes: thermal detectors or quantum detectors. Within each class there are a number of specific types which show differences in the detection mechanism, the readout method or the physical structure of the detector. Table 2 lists some of the more commonly used types of detectors in each classification. It also lists under 'other detectors' some of the available detectors of optical radiation that do not provide an electrical output. Definitions and operating principles are given below for the various detector types.

Table 2   Detection of optical radiation

Quantum detectors	Photoemissive
	Photoconductive
	Photovoltaic
	Photojunction
	Photon drag
Thermal detectors	Thermistor
	Bolometer
	Thermocouple/thermopile
	Golay cell
	Pyroelectric
Other detectors	Photochemical
	Photoplastic
	Phosphors
	Photochromic

Detailed comparisons of various detector classes and types require the definition and evaluation of performance parameters; these parameters are defined later. However, some initial comparisons beween the thermal and quantum detector classes are given in Table 3.

Quantum detectors

The fundamental principle of quantum detection is illustrated in Fig. 6 by considering the allowed energy states in a semiconductor material as related to the energy of photons incident upon that material. A minimum energy is required of absorbed photons if charge carriers are to be released from their bound states and made available for conduction. These free charge carriers change the electrical properties of the semiconductor material.

A variety of energy bandgap values are possible by selecting single, binary, and tertiary intrinsic semiconductor materials or by designing extrinsic semiconductor materials using various impurity elements and dopant levels. Since photon energy is inversely proportional to wavelength, the minimum photon energy required to produce a charge carrier (as indicated in Fig. 6) determines the maximum wavelength to which a quantum detector will respond. Response to shorter wavelengths (higher energy photons) is limited by the ability of the detector material to absorb (stop) the photons.

Absorption efficiency for most materials decreases as photon energy increases. The result of these two limits is that a quantum detector responds

Table 3   Comparison of thermal and quantum detectors

	Quantum	Thermal
Spectral response	Narrow band	Broadband
Sensitivity	High	Medium
Time response	Fast*	Slow*
Cooling required	Yes (some)	No
Response uniformity	Good	Fair

* Some thermal detector designs can achieve fast time response. If the quantum detector is cooled its time response is slower

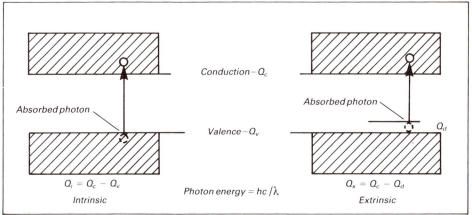

Fig. 6 Fundamental principle of quantum detection

only to a relatively narrow range of wavelengths (energies) and is characterised as having a narrow-band spectral response.

The following descriptive definitions are given for certain types of quantum detectors.

● *Photoemissive.* Photoemissive materials have a low surface work function so that absorbed photons produce free electrons with sufficient energy to escape the material. Thus, detection consists of the conversion of photons to emitted electrons. The photoemitter is typically enclosed in a vacuum tube with various schemes for collecting and measuring the emitted electrons. Commonly used photoemitter materials require relatively high-energy photons in order to exceed the surface work function and thus respond to visible or shorter wavelengths.

● *Photoconductive.* The conductance of a semiconductor material is increased as absorbed photons produce free charge carriers in the material. The change in conductance is measured by applying an external bias voltage across the detector and measuring the change in current through the detector. Response to various wavelengths is governed by the energy levels in the semiconductor material as indicated in Fig. 6.

● *Photovoltaic.* Photons absorbed at a *p–n* junction produce hole–electron pairs that change the junction barrier potential. Thus, the change in the open-circuit output voltage of the detector is a measure of the optical radiation. The photovoltaic detector does not require an external bias voltage. As with the photoconductor, the spectral sensitivity of the photovoltaic detector is governed by the available energy levels in the semiconductor material.

● *Photojunction.* Photon absorption by diode or transistor configurations causes a change in their electrical properties. Photojunctions may be photoconductive or photovoltaic.

● *Photon drag.* The momentum of absorbed photons is transferred to free charge carriers in a semiconductor material.

Thermal detectors

Thermal detectors absorb photons or radiant energy over a relatively broad band of optical wavelengths. This absorbed energy causes a rise in the temperature of the detector material, which is chosen to have a measurable temperature-dependent parameter. One of the simpler thermal detectors is a thermistor, which has a temperature-dependent resistance. Thermal detectors are usually coated with materials that increase their absorptance for enhanced sensitivity. The time response of a thermal detector is limited by how fast it can change temperature, which is related to its heat capacity, its physical dimensions, and the properties of the required heat sink. Descriptive definitions of a number of thermal detectors are given below.

● *Thermistor.* A thermistor is a temperature-sensitive resistor. Good thermistor materials are those that have a high temperature coefficient of resistance. Typical thermistor materials include metallic oxides and semiconductors. These detectors have high resistivity so that heating caused by current flow in the detector is minimal. Metals also exhibit a temperature-dependent resistivity; however, as inherently low-resistivity materials they will have considerable heat generated from current flow. This current-generated heat energy cannot be easily distinguished from the optically induced heat. Furthermore, the stability of the metallic thermistor is poor. Thermistors have been fabricated using pastes, with thin-film techniques, and as individual semiconductor elements.

● *Bolometer.* A bolometer is an instrument using a temperature-sensitive resistor (a thermistor or metallic detector). The output can be measured as the voltage across a load resistor in series with the thermistor and a bias voltage. Or, a more sensitive instrument is obtained if the thermistor is connected in a Wheatstone bridge circuit.

● *Thermocouple/thermopile.* An emf is generated between two junctions of dissimilar materials if the two junctions are at different temperatures. This is the Seebeck effect and is the basic operating principle for a thermocouple. The materials in the thermocouple may be metals or semiconductors, with semiconductors being more sensitive. The use of a thermocouple as an optical radiation detector requires that one junction be exposed to the radiation and that the other junction be shielded from the radiation. Since the output of a thermocouple is typically very small, the sensitivity can be increased by connecting several thermocouples in series. This series arrangement of thermocouples is called a thermopile.

● *Pyroelectric.* The pyroelectric detector consists of a thin slab of ferroelectric material sandwiched in between two electrodes. Ferroelectric materials exhibit a spontaneous polarisation that is temperature sensitive. Thus, the pyroelectric detector is a temperature-sensitive capacitor. The pyroelectric detector does not require a bias voltage and it produces an output voltage that is proportional to the time rate of change of the temperature. This makes possible a fast time response for the pyroelectric detector.

● *Golay cell.* The Golay cell is a pneumatic device and is essentially a gas thermometer. Optical radiation is focused onto a small cell containing a gas. Absorbed energy increases the temperature and pressure of the gas. The increased gas pressure expands a bellows to deflect a mirror and give an optical readout. The Golay cell is a very sensitive optical detector; however, it is also very fragile and does not find use outside the laboratory environment.

Detector principles

Most quantum detectors operate as photoemitters, photoconductors, or photovoltaic devices. The fundamental principles of these three types of quantum detectors are given below along with the principle of the pyroelectric detector.

Photoconductive. The principle of photoconductivity is illustrated by considering the simple model for a detector as shown in Fig. 7. The current I consists of the sum of a quiescent term and a photoinduced term, $I = I_{dc} + I_s$. The photoinduced current is given by $I_s + dQ_s/dt$, where dQ_s is the charge induced in the detector by the incident photon flux E_s given by:

$$dQ_s = E_s \eta (wl) \tau q \qquad (15)$$

where η is the detector quantum efficiency (charge carriers per photon), τ is the charge carrier mean lifetime (s), q is the electronic charge (C), and wl is the irradiated surface area (m^2).

The transit time dt of these charge carriers across the detector width l is on the average given by:

$$dt = \left| \frac{\mathbf{l}}{\mathbf{v}} \right| = \frac{l}{\mu |E|} = \frac{l^2}{\mu V_{bias}} \qquad (16)$$

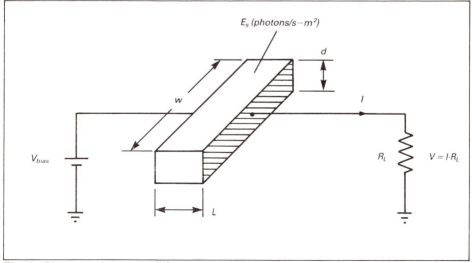

Fig. 7 Photoconductor model

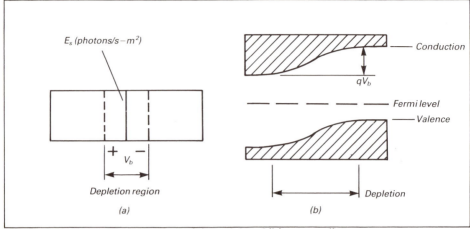

Fig. 8   The photovoltaic detector: (a) model and (b) energy diagram

where v is the charge carrier velocity, μ is mobility (m²/V s), and E is the electric field intensity vector across the dimension l. The photoinduced conductance in the detector is directly proportional to the incident photon flux E_s:

$$\sigma_s = E_s \eta \tau q \mu / d \qquad (17)$$

Photovoltaic. The photovoltaic detector is the only quantum detector not requiring a bias voltage supply. Its contruction commonly consists of a semiconductor p–n junction as shown in the model in Fig. 8. In the absence of photon flux, the junction barrier potential has the polarity shown in and a magnitude given by:

$$V_b = \frac{kT}{q} \ln\left(\frac{n_n}{n_p}\right) \qquad (18)$$

where n_n and n_p are the free electron densities in the n material and the p material, respectively.

Now consider an incident photon flux on the p–n junction that produces additional hole–electron pairs. Additional electrons freed in the n material are then available to fill holes in the depletion region of the p material. The net result is an increase in n_n and a decrease in n_p. If these changes are given by δn_n and δn_p, respectively, the junction barrier potential becomes:

$$V_b = \frac{kT}{q} \ln\left(\frac{n_n + \delta n_n}{n_p + \delta n_p}\right) \qquad (19)$$

The changes in the number of charge carriers in the n and p materials should be approximately equal in magnitude and given by:

$$\delta n_n = \delta n_p = \frac{\eta E_s}{d} \quad \text{(electrons/m}^3\text{)} \qquad (20)$$

where d is the detector depth, η is the quantum efficiency, and E_s is the photon flux on the junction area. The change in V_b for a change in photon flux gives the photoinduced voltage in the detector. Using the approximation that the ratio of δn_n to n_n is small gives a linear relationship between the photoinduced voltage and the incident photon flux:

$$\frac{dV_b}{dE_s} \approx \frac{kT\eta(n_n + n_p)}{qdn_n n_p} \qquad (21)$$

Photoemissive. Good photoemissive materials are those with low electron affinity or surface work function. The basic principle is illustrated in Fig. 9 along with a diagram showing the typical use for photoemitters as optical detectors. An absorbed photon with energy hv joules imparts sufficient energy to release an electron from the material. For an electron initially in the valence band, its energy after leaving the material is given by:

$$Q = hv - (Q_e - Q_v)\,(\text{J}) \qquad (22)$$

where Q_e is the minimum energy an electron must have to escape the surface of the material. In a detector configuration, the photoemitter is enclosed in a vacuum envelope. The emitted electrons are then accelerated

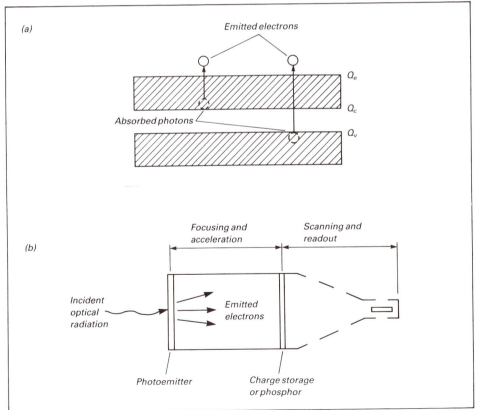

Fig. 9   Photoemissive detectors: (a) energy level diagram and (b) typical use

and focused electronically onto either a charge storage target (for signal-generating image tubes) or a phosphor screen (for image intensifiers). For signal-generating image tubes, the image-dependent stored charge is read out by a scanning electron beam.

Because the photon energy must be high enough to free electrons from the emitter material, most available photoemitters respond to only visible and shorter wavelengths, with limited response in the near infrared. A primary nonimaging application for photoemitters is in photomultiplier tubes, which are high-sensitivity optical detectors that use several stages of internal gain. Older television camera tubes such as the image orthicon and image isocon used photoemitters as the photodetection principle.

Pyroelectric. Since pyroelectric detectors are a subclass of thermal detectors, they typically exhibit a broad spectral response. However, they are different from other thermal detectors in that their response is proportional to the time rate of change of temperature, not the temepature itself. This property allows a much faster time response for pyroelectric detectors as compared with other thermal detectors.

Pyroelectric materials are ferroelectric crystals which show spontaneous polarisation along one axis (in the absence of an electric field). Furthermore, this spontaneous polarisation is dependent on the temperature of the crystal and is characterised by a pyroelectric coefficient p with units of C/m^2K. For a polarisation P (C/m^2) and temperature T (K), the pyroelectric coefficient is given by:

$$p = dP/dT \qquad (23)$$

A model for the pyroelectric detector is shown in Fig. 10 using face electrodes; that is, the irradiated face of the detector is also the electrode face. Edge electrodes are an alternate configuration. For both face and edge electrodes if the electrodes are such that the polarisation vector is perpendicular to the electrode surface then the configuration is that of a temperature-sensitive capacitor.

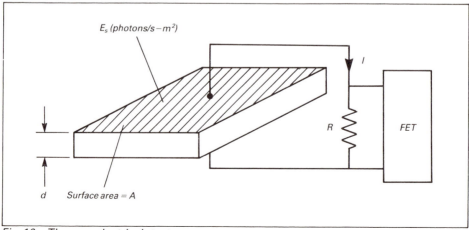

Fig. 10   The pyroelectric detector

In the absence of optical radiation the spontaneous polarisation P causes a charge $q = PA$ (C) on the electrode surfaces. This charge is neutralised by current flow through the resistor R. If the detector is open circuited and leakage currents are neglected, then the detector will have a voltage across its electrodes of $V = q/C = qd/\varepsilon A$ (V).

Absorbed radiant flux E_s raises the detector temperature by an amount δT causing a change in the crystal polarisation of δP. This change in polarisation gives rise to a surface charge, due to the pyroelectric effect, on the electrodes of $\delta q = A \delta P = Ap \delta T$.

With the load resistor in place, δq is neutralised by a current flow:

$$i = \frac{\delta q}{\delta t} = Ap \left(\frac{\delta T}{\delta t} \right) \tag{24}$$

Thus, the photoinduced current in the pyroelectric detector is proportional to the time derivative of the temperature. This allows the pyroelectric detector to have a faster time response than other thermal detectors. The photoinduced input voltage to the field effect transistor in Fig. 10 is given by $\delta V = RpA(\delta T/\delta t)$. Thus, the response is zero for a constant detector temperature. Furthermore, since the detector temperature is proportional to the radiant flux E_s, the pyroelectric detector requires that the optical radiant flux be chopped (alternately blocked and passed) or that scanning be provided. The exact dependence of detector temperature on radiant flux requires solution of the heat equation for a specific detector configuration and model.

Detector performance parameters

Two primary measures of detector performance are its sensitivity and noise immunity. Sensitivity is used here in a general sense such that higher sensitivity is better; however, for optical detectors there is no specific performance parameter called 'sensitivity'. Noise is usually the limiting factor in determining how small a signal change can be measured by the detector. Two parameters are defined that provide a measure of the detector's immunity to noise. Fundamental detector performance is quantified by parameters and functions that relate the sensitivity and noise of the detector to wavelength and modulating frequency of the optical radiation. The following performance measures are generally applicable to all types of optical detectors. In addition, certain specific detector types will have additional performance measures that result from either their construction or the mechanism for reading out the signal. Selected special cases that are of primary interest to electro-optics include focal plane arrays and image tubes.

Responsivity

Responsivity is a measure of the detector's ability to convert optical flux into a measurable output parameter (typically volts or amperes). In one sense it gives the gain of the detector. To show dependence on wavelength we define spectral responsivity as the output signal for an input optical flux in a differential wavelength band $d\lambda$ centred about wavelength λ.

$$R_\lambda = \frac{s_\lambda}{\phi_\lambda d\lambda} \text{ (V/W or A/W)} \tag{25}$$

where s_λ is the detector output in volts or amperes for an incidental optical flux of $\phi_\lambda\, d\lambda$ (W). Variations on the units include representing the optical flux in photons/second, using a normalised response (with a maximum value of unity) versus wavelength, or plotting quantum efficiency versus wavelength. The exact shape of the spectral responsivity curve will not be the same for different choices of units. This is illustrated by the fact that 1W of optical flux converts to a photon flux that is wavelength dependent.

The total responsivity of an optical detector is defined as the total output signal divided by the total radiant flux incident on the detector and is given by:

$$R = \frac{\int_0^\infty s_\lambda d\lambda}{\int_0^\infty \phi_\lambda d\lambda} = \frac{\int_0^\infty R_\lambda \phi_\lambda d\lambda}{\int_0^\infty \phi_\lambda d\lambda} \tag{26}$$

From this result it is clear that total responsivity depends not only on the spectral response properties of the detector as given by R_λ. Thus, realistic comparison of various detectors is best accomplished by using the spectral responsivity or by computing a total responsivity for an assumed constant source over the wavelength range of interest. Total responsivity stated as a measurement value for a a particular detector has little meaning unless the spectral properties of the source used for that measurement are known. It is common practice to use a black-body source at a given temperature as the source for such measurements.

In addition to the dependence on wavelength as given above, responsivity will show a dependence on temporal frequency as well. For most optical detectors this temporal dependence exhibits a lowpass characteristic and can be described by:

$$R_\lambda (f) = R_\lambda (0)/[1 + (f\tau)^2] \tag{27}$$

where τ is the responsive time constant for the detector.

Noise equivalent power

It was shown above that responsivity is a measure of the ability of a detector to convert optical flux to an electrical signal. A large value of responsivity is good because it produces a large output signal for a given radiant flux; however, responsivity gives no information as to how small the optical flux can be and still be detected. In other words it gives no information about the noise limitation of the detector. The noise equivalent power (NEP) is a performance parameter that does give a measure of the minimum detectable flux and is defined as the level of incident radiant flux that produces an output signal-to-noise ratio of unity. Noise equivalent power has units of watts and is given by the relationship:

$$s_\lambda = R_\lambda \, \text{NEP}_\lambda = N_{\text{rms}} \tag{28}$$

where $\text{NEP}_\lambda = \phi_\lambda \, d\lambda$ is the incident flux at wavelength λ required to produce a signal-to-rms ratio of unity of an electronics bandwidth of $\triangle f$ (Hz). Further quantification of NEP requires that the limiting specific detector noise be identified and characterised. If the optical flux is modulated so that $1/f$ noise is not a problem, then the limiting noise is typically either thermal noise in the detector/preamp combination or fluctuation in the number of charge carriers in quantum detectors. For thermal detectors the limiting noise is typically temperature noise in the detector. In addition to the limiting noise, NEP clearly depends on the spectral properties of the incident flux, the spectral responsivity of the detector, and the electronics bandwidth. Bandwidth depedence is usually eliminated by normalising to a 1Hz bandwidth at a given modulating frequency. Furthermore, dependence on properties of the optical source are eliminated by using the spectral NEP to characterise the detector.

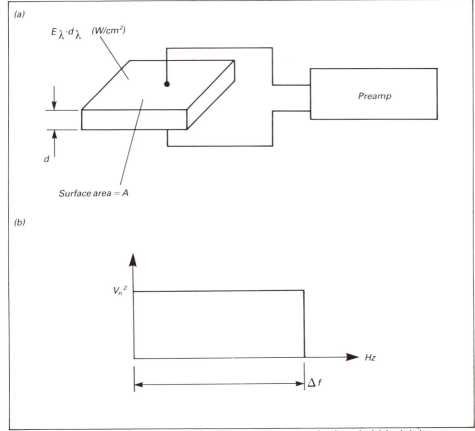

Fig. 11   Noise dependence on detector area and electronics bandwidth: (a) detector model and (b) noise power spectrum

Detectivity and specific detectivity

Detectivity is defined as the reciprocal of NEP and has the psychological advantage that it increases for improved detector performance, whereas NEP decreases.

$$D = \frac{1}{\text{NEP}} \; ; \; D_\lambda = \frac{1}{\text{NEP}_\lambda} = \frac{R_\lambda}{N_{\text{rms}}} \; (\text{W}^{-1}) \tag{29}$$

The values for NEP and detectivity depend on the noise bandwidth and the detector area. The dependence is illustrated by considering noise due to fluctuation of charge carriers in the detector. Consider the detector/preamplifier model shown in Fig. 11. For shot noise processes such as that represented by the fluctuation in charge carriers, the rms noise is equal to the square root of the number of charge carriers. At a given wavelength, the incident power on the detector in terms of photon flux is given by:

$$\phi_p = \frac{\phi_\lambda d_\lambda A}{hc/\lambda} \; (\text{photons/s}) \tag{30}$$

For an quantum efficiency of η and a mean charge carrier lifetime of τ (s), the number of photoinduced charge carriers is given by:

$$N = \phi_p \eta \tau = \frac{\lambda \eta \tau A \phi_\lambda \, d\lambda}{hc} \; (\text{charge carriers}) \tag{31}$$

where N is the mean number of photoinduced charge carriers. The rms fluctuation is given by the square root of N, which is proportional to the rms noise voltage V_n at any frequency, as shown in Fig. 11. If we integrate over all frequencies, the total mean noise power is given by:

$$V_n^2 \, \triangle f = \frac{\lambda \eta \tau A \phi_\lambda \, d\lambda \, \triangle f}{hc} \tag{32}$$

and the total rms noise for use in Eqn. (29) is:

$$N_{rms} = \left(\frac{\lambda \eta \tau A \phi_\lambda \, d\lambda}{hc} \right)^{1/2} (A \, \triangle f)^{1/2} \tag{33}$$

The point is that NEP is proportional to the square root of detector area and electronics bandwidth, and detectivity is inversely proportional to these same quantities. Other parameters in the result in Eqn. (33) are either wavelength dependent or dependent on the detector material (e.g. quantum efficiency η and carrier lifetime τ) and should be fixed for a particular detector material. Detector area and electronics bandwidth, however, are expected to vary with the application. Rather than have to specify bandwidth and detector area, a normalised detectivity is defined. Specific spectral detectivity D^* is defined as:

$$D_\lambda^* = (A \,\triangle f)^{\frac{1}{2}} \, D\lambda = \frac{(A \,\triangle f)^{\frac{1}{2}}}{\mathrm{NEP}_\lambda} \quad (\mathrm{m \, Hz^{1/2}/W}) \qquad (34)$$

Units for D^* are more commonly given as cm $\mathrm{Hz}^{1/2}/\mathrm{W}$. It is proportional to spectral responsivity and shows a dependence on temporal frequency given typically by:

$$D_\lambda^*(f) = \frac{D_\lambda^*(0)}{[1 + (f\tau_d)^2]^{\frac{1}{2}}} \qquad (35)$$

where τ_d is called the detective time constant. Furthermore, the relationship between responsivity and specific detectivity is given by:

$$D_\lambda^* = (A \,\triangle f)^{\frac{1}{2}} R_\lambda / N_{\mathrm{rms}} \qquad (36)$$

Frequency response

Conversion of radiant flux to an electrical signal shows time constraints because of a number of limiting factors in the detection process. These include the following phenomena:

● *Charge carrier lifetime* – The mean lifetime of photogenerated charge carriers in quantum detectors.

● *Charge transport times* – Charge carriers generated in the detector require a finite amount of time to move through the detector material. This phenomenon is essentially the detector capacitance and is dependent on detector size and charge carrier mobilities.

● *Thermal capacitance* – Of primary importance for thermal detectors this phenomenon results from the fact that detector materials require a finite amount of time to change temperature in response to an incident radiation flux.

● *Electronics* – The preamplifier and amplifier will exhibit finite rise and fall time characteristics.

● *Modulation of the radiant flux* – The radiant flux is modulated either by scanning or optical chopping to enhance noise reduction, to cover the field of view, or as a necessity for certain detector types (e.g. the pyroelectric detector).

The result is a temporal dependence for all the above-defined performance parameters. For a given detector, the internal time-dependent parameters such as charge carrier lifetime, mobility, and detector geometry are considered to be fixed. Choices can still be made for modulation frequency and electronics bandwidth. Detector capacitance (charge transport time) is dependent on internal parameters and on the level of incident radiant flux. The overall response time for a detector depends on detector temperature, design parameters, and its radiation environment. Thus, it is imperative that any stated time response for a detector be evaluated relative to the conditions under which it was measured.

Temperature dependence

The performance of an electro-optical detector depends on the temperature of the detector itself and on the temperature of its surroundings. The detector temperature strongly affects the thermal noise generated within the detector and thus affects the value of NEP and detectivity. For quantum detectors in particular the performance is also affected by the temperature of its surroundings and any object in its field of view. Photons from the detector's surroundings increase the radiation noise in the detector and thus affect NEP and detectivity.

Analyses that show the dependence of D^* on the detector temperature and the incident photon flux indicate that the detector performance is typically limited by one of these two sources of noise. For temperature-limited operation the thermal noise in the detector is dominant; whereas, for background-limited operation the radiation noise is dominant.

A number of specific noise processes can contribute to the total NEP in a detector. All eventually come to be expressed in terms of a fluctuation in the number of charge carriers in a semiconductor detector. For thermal detectors the result is typically expressed as a fluctuation in either voltage or current. The two dominant noise sources in many semiconductor detectors are thermal noise and background (or radiation) noise. A system using these detectors in most cases will be designed to be background-noise limited if possible.

Performance limits due to the detector temperature are typically shown as the dependence of D^* on detector temperature. Many quantum detectors exhibit a threshold effect with respect to temperature; that is, if the temperature of the detector rises above a threshold value then D^* drops off almost instantaneously.

Background-noise-limited operation is described in terms of the dependence of D^* on the level of incident photon flux, recognising that noise is proportional to the square root of that flux. Background-limited D^* is given by [4,5]:

$$D^*_{\text{blip}} = \frac{\lambda}{2hc} \frac{\eta^{1/2}}{(E_B)^{1/2}} \tag{37}$$

where η is the detective quantum efficiency and E_B is the incident photon flux on the detector from all sources. These include the background, optical path, stray sources, the objects in the field of view, the optical elements, and the sensor housing. Obviously a good sensor design minimises the effect of these sources so that only those objects in the field of view have much effect; that is, the optical elements, the optical path, and the scene within the detectors field of view.

Spectral detectivity curves given in the literature are for a specific detector operating temperature. Furthermore, the curves are for a 2π steradian field of view and a 295K background temperature. The net result is that the detector has a very large E_B incident upon it. For this reason actual operating characteristics for the detector enclosed in a well-designed sensor will be far better. Increases in D^* of 3–4 orders of magnitude can be

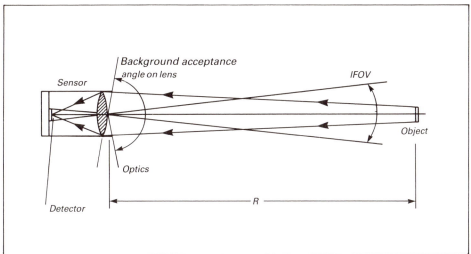

Fig. 12   Background photon flux on a detector

achieved. Irradiance incident on the lens from an atmospheric path and radiant exitance from the lens itself will have essentially a Lambertian distribution so that the detector will actually receive photons from the entire viewing cone of the lens/detector arrangement as shown in Fig. 12. The object is smaller than the detector instantaneous field of view so that it will be imaged entirely on the detector. Thus, the photon flux on the detector from the object is the flux collected by the lens. All these sources may be quantified as:

$$E_B = \int_\lambda \Bigg((M_a + M_o)(NA^2) + \frac{M_{\mathrm{obj}} A_{\mathrm{obj}} D^2}{4R^2} \Bigg)\, d\lambda \qquad (38)$$

where M_a and M_o are spectral radiant exitance from the path and optics, respectively, M_{obj} and A_{obj} are the radiant exitance and area of the object. Spectral dependence for the radiant exitance terms is left implicit and does exist. Typically, λ in Eqn. (37) is taken as the centre wavelength in the spectral band of interest.

Sensor types, constraints, and objectives

It is difficult to describe the elements of a general electro-optical sensor system because there is such a wide range of possible types and configurations. However, Fig. 13 shows the major elements that may be included and that will impact the design of the system. The system may be active or passive. In the case of an active system, an optical source irradiates the scene with selected wavelengths. The sensor is then designed to be maximally sensitive to the chosen wavelengths. It responds to reflected radiant flux from the scene. A passive system, on the other hand, depends on radiant energy emitted by the scene or reflected from its ambient environment (e.g. reflected sunlight). The sensor may be imaging or non-imaging and may be sensitive in one or more spectral bands.

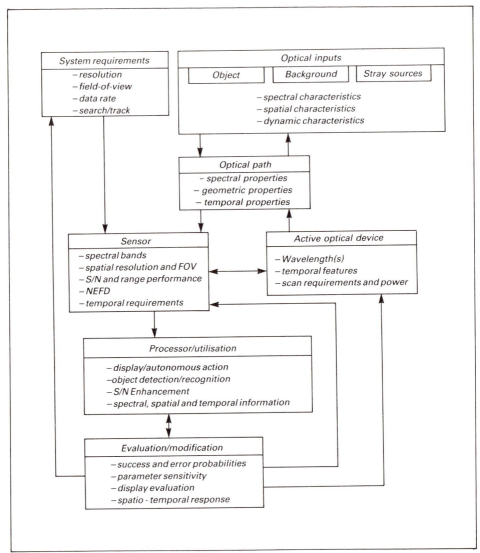

Fig. 13   Typical elements of an electro-optical system

The actual components of the sensor are grouped into three categories: (1) optics, including all the optical elements (mirrors, lenses, filters) and the physical arrangement of those elements within a sensor housing; (2) photosensor, the detector subassembly including the detectors and first-level preamplification as well as a mechanism for cooling when required; and (3) electronics, including scanning circuitry and signal electronics. Performance parameters for these sub-elements are combined with parametric descriptions of the optical signature and optical path to develop expressions for signal-to-noise ratio and resolution.

In order to begin the design of any electro-optical sensor system, a number of questions have to be answered. The answers to these questions form the system constraints that make it possible to begin assigning values

Table 4 System constraints

Objectives
  (1) Imaging or non-imaging
  (2) Detection/recognition/identification of objects
  (3) Real-time operation
  (4) System longevity, size, weight, maintenance
  (5) Object tracking

Scenario
  (1) Controlled or natural environment
  (2) Ambient environment characteristics
    – sun position
    – number and location of ambient sources
    – ambient temperature
    – optical path (atmosphere, space, controlled)
  (3) Object characteristics
    – spectral signature
    – temperature, size, shape
    – number and location of objects
  (4) Background characteristics
    – terrain, atmosphere, space, controlled
    – spectral signatures
  (5) Overall geometrics
    – maximum and minimum distance to object(s)
    – range and size relationships
    – required resolution and total field of view
  (6) Sensor/scene dynamics
    – relative motion and velocities
    – required motion and velocities
    – required update rate for processor

---

to parameters in the performance equations. For example, Table 4 lists a number of constraints in two categories. Essentially, it must first be decided what the objectives are for the system and under what scenario the system is to function. Again, it is evident from looking at Table 4 that the range of questions to be answered that cover all possible sensor applications is quite varied. Once these questions have been answered, however, the range of possibilities narrows very quickly and the problem of sensor design becomes manageable.

The next step is to use the known system constraints to select parameters for the actual sensor. For example, one of the first sensor decisions is what are spectral bands (one or more) to which the sensor must respond. Based on object, background, and ambient radiant properties as well as the spectral transmittance properties of the optical path, choose one or more spectral bands that exhibit (1) strong object signature, (2) minimum interfering background radiation, (3) minimum interfering stray ambient sources, and (4) low attenuation by the optical path.

Within the selected candidate spectral bands, a further narrowing down of options is dependent on the availability of detectors that are sensitive to those wavelengths and that provide geometries (detector size, number in array, cost) that meet the requirements of spatial resolution.

The constraints of imaging versus non-imaging, real-time operation, and the total field-of-view requirement impact the choice of detector as well. For example, an imaging sensor that must cover a wide field of view with high spatial resolution can meet these requirements more easily with a focal plane array containing many detectors. Objectives for the sensor may include the detection, recognition, or identification of objects within the field of view. These increasing levels of difficulty will require more detailed information from the sensor. For example, increasingly difficult processing tasks will most often require higher resolution, higher signal-to-noise ratio, and perhaps multispectral images. These choices again impact strongly the selection of detectors, and of course the availability of detectors will impact whether the objectives can be met.

The point made by the above discussion is that design of a sensor is an iterative process that goes through many iterations in an attempt to match objectives with capability. Invariably, a detailed trade-off analysis is performed to obtain the best design. For applications that impose stringent demands (pressing the state-of-the-art in available technology) the trade-off analysis becomes something of an art and requires creativity to achieve the goals. Other applications are such that goals are easily achieved with commercially available sensors (such as television cameras – vidicon or solid state) and the performance analysis is relatively straightforward.

The only way to learn sensor analysis and design is by example and experience. As an aid in presenting the design technique, a detailed development of two diverse sensor design examples is given to show the methodologies.

The two examples chosen for analysis represent two levels of complexity. The first is a simple sensor whose objective is detection of an object against

Table 5   Example 1 system constraints/objectives

Objectives
- non-imaging
- detect and track multiple objects in field of view
- tracking update once per second
- zero maintenance, long lifetime
- distinguish objects from stars

Scenario
Ambient environment
- space environment (sensor and objects)
- space optical path
- potential ambient sources (sun, earth, moon, stars)
Object characteristics
- temperature: 300K
- spectral properties: grey-body ($\varepsilon \times 0.8$)
- size: 1-5m^2 cross-sectional area
- number of objects: max of 10 in FOV
Background characteristics
- space(no signature)
- sensor never pointed at sun or earth
- some stars may be detectable
- constant attitude with respect to earth
Overall geometries
- range 500–5000km
- resolution required: 1km max at 5000km range
- total field of view: 10° × 10°
Sensor/scene dynamics
- sensor and objects at about same orbital altitude
- relative velocities less than 1km/s

---

a space background. Such a sensor would be useful for satellite tracking. Its complexity is increased slightly by the requirement to distinguish between stars and objects (satellites) in the field of view. The second sensor is more complex with a requirement for producing a high-resolution image with the aid of an active irradiator so that object identification is possible in a relatively complex scene. Applications for this type of sensor include robotics and industrial automation. The environment may be controlled (as with assembly line operations) or complex (as with a mobile robot). System constraints and objectives for the two examples are given in Table 5 for the simple system and in Table 6 for the complex system.

Performance evaluation: nonimaging sensor

From the system objectives and constraints listed in Table 5 the sensor design proceeds by identifying certain candidate parameter values. Based on the requirement for a maximum resolution cell size of 1km (assumed to be $1 \times 1 \text{km}^2$) at a range of 5000km, the instantaneous field of view (IFOV) in each direction for a single detector must be no more than:

$$\text{IFOV} \leqslant 1/5000 \times 0.2 \text{mrad} \tag{39}$$

Since the total field of view in each direction is 10° (or approximately 175mrad), then the total field of view contains 175/0.2 or 875×875 resolution cells. To provide the stated resolution over the total field of view requires either an array of 875×875 detectors or the scanning of a few detectors over the field of view by optical or mechanical means. The best choice depends on several considerations including the desired wavelength region for the sensor operation and the availability of large-scale arrays that operate in that spectral band.

Table 6   Example 2 system constraints/objectives

Objectives
  – high-resolution imaging (0.5mm on object)
  – object identification in real-time
  – active irradiation (visible)

Scenario
  Ambient environment
      – controlled, short range
      – interfering sources minimised
  Object characteristics
      – temperature: ambient
      – three distinct object classes: shape or size difference
      – different spectral properties for three classes
  Background characteristics
      – conveyor belt: low optical signature
  Overall geometries
      – sensor to object distance: 50–150cm
      – object sizes: three, 3 and 4cm diameter approx.
      – conveyor belt: 15cm width
      – objects never lay on top of each other
  Sensor/scene dynamics
      – conveyor velocity: 10 cm/s
      – average of 10,000 objects/h to be identified

A choice of one or more spectral bands is based on the spectral properties of the objects as compared with the spectral properties of stray objects that may be in the field of view. For the constraints in Table 5, the objects will produce an optical signature that consists largely of an emitted component (from a 300K grey-body with emissivity of 0.8) with a smaller reflected component (reflectance = 0.2) from strong ambient sources (sun and earth). Keeping in mind that the sensor must distinguish between objects (at 300K) and stars (at several 1000°C) in its field of view, an approach that offers discrimination on the basis of temperature is to choose two spectral bands: one in the infrared region near 8–14µm (the 300K object has peak exitance at about 10µm) and one in the visible region near 0.4–0.7µm. Stars in the field of view will produce a significant optical signature in both spectral bands (stars are typically modelled as black-body sources); whereas the objects (at a temperature of 300K) produce essentially no radiant energy in the visible region.

An alternative choice of spectral bands for which there are workable quantum detectors is 8–14 and 3–5µm. Discrimination of stars is still possible using the fact that the ratio of radiant energies in two spectral bands is a monotonic function of temperature. In particular, the ratio:

$$\rho(T) = \frac{\int\limits_{\triangle\lambda 1} M_\lambda(T)\, d\lambda}{\int\limits_{\triangle\lambda 2} M_\lambda(T)\, d\lambda} \tag{40}$$

can be computed as a function of temperature for various wavelength regions and used as a first-level temperature discriminant.

Quantum detectors that are sensitive in the 3–14µm range are available but not in arrays of 875×875 elements. Therefore, the approach for the sensor design will be to use one or a few detectors and provide coverage of the field of view by opto-mechanical scanning. Linear arrays of up to 2048 elements are available for certain types of detectors. The use of a linear array of 1024 detectors would make possible a simple linear scan for this example. Clearly a linear optomechanical scan is simpler than a two-dimensional scan.

The resulting sensor will operate in a pulse detection mode. Furthermore, its performance is described in part by the range equation developed by Genoud[6]:

$$\frac{S}{N} = \frac{M_{\triangle\lambda}\varepsilon A_s T_{at_o} D^2\, D^*}{4R^2} \left(\frac{t_d V}{A_d} \right)^{1/2} \tag{41}$$

In terms of the detector configuration there are still some choices to be made. The simplest detector configuration is one that has a single detector whose field of view is scanned over the total field of view (TFOV) optically or mechanically. In order to meet the required data rate of once per second, the complete scan must be accomplished in 1s. With this configuration the pulse dwell time on the detector for any object is given by:

$$t_d = \frac{\text{IFOV}}{\text{TFOV}} = \frac{(0.0002)^2}{(0.175)^2} = 1.3\mu s \tag{42}$$

where the TFOV must be scanned in 1s and one resolution cell is scanned in t_d seconds. A pulse dwell time as small as 1.3μs requires an electronics bandwidth proportional to the inverse of dwell time, or approximately 770kHz. This large bandwidth significantly increases noise in the detector and degrades sensor performance.

An alternative detector configuration that may reduce the bandwidth requirement is one that has a linear array of detectors covering the total field of view in one direction (875 detectors). Scanning is then required in only one dimension and the pulse dwell time becomes:

$$t_d = \frac{0.0002}{0.175} = 1.14\text{ms} \tag{43}$$

Thus, a more reasonable bandwidth requirement is achieved at the expense of adding more detectors and associated electronics. It is emphasised here that the saving in bandwidth assumes each detector has its own readout electronics. In arrays where readout is by sequential electronic scanning of the array, the bandwidth must again be large enough to provide readout of a given detector in 1.3μs. The final choice of detector configuration is best made after some performance calculations to see which if either configuration satisfies the system objectives. Remember also that the requirement for measurement in two spectral bands will add additional complexity to the detector configuration.

Looking at other parameters in Eqn. (41), since the system is operating in a space environment, the path transmittance T_a is unity; and from Genoud's article the electronic gain factor V has a range of 1–2 so that a nominal value of 1.2 may be chosen without detailed justification at this point. The parameters A_d and D^* are determined to some extent by the specific detector type to be used and the manufacturing technology for that detector

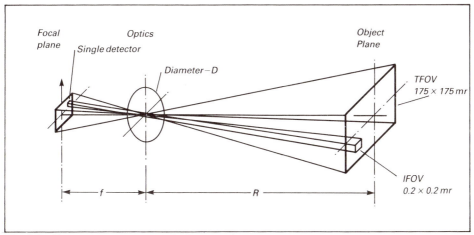

Fig. 14   Sensor system example 1 geometry

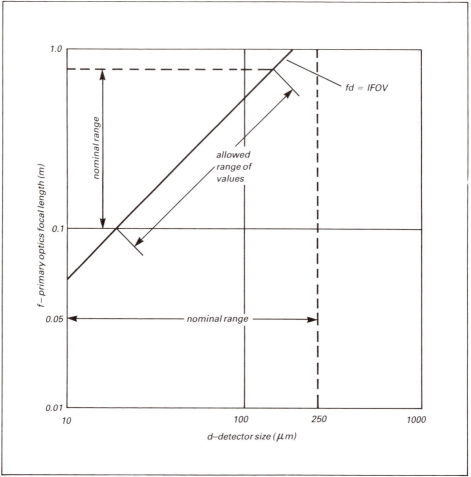

Fig. 15   Constraints on detector size and optics size

material. Furthermore, D^* will be limited in value by either thermal noise in the detector or background (radiation) noise from the operating environment. For the space system in this example the only significant source of background radiation striking the detector is from the optical elements of the sensor itself. Since optical elements typically have very low emissivity and because of the low space temperatures, the background radiation on the detector will be low. If the detector is cooled to reduce thermal noise, then, beacuse of the low radiation noise, it is expected that the obtainable D^* will be very high. Published spectral D^* curves for detectors are typically for a recommended detector operating temperature and for a 2π steradian field of view against a 295 or 300K background. One approach for determining D^* is to assume it is operating at a temperature where thermal noise is insignificant and to calculate the background-limited D_{blip} as described earlier.

As a next step in tying down acceptable values for parameters, consider the constraints on resolution as indicated in Fig. 14. From the requirements stated in Table 5, the IFOV at maximum range, 5000km, is as shown in the figure. From simple geometric optics, the detector linear dimension and the optics focal length must satisfy the same relationship, that is:

$$d/f = \text{IFOV} = w\,R_{\text{max}} = 0.0002 \tag{44}$$

This parametric relationship is significant because there are practical limits on both the detector size and the optics size. Recognising also that optics focal length is related to optics diameter by $f = (f/\#)D$ and that S/N is proportional to D^2, it becomes clear that resolution parmeters cannot be chosen independently of other considerations. Fig. 15 shows a plot of optics focal length in metres versus detector size in micrometres. The curve is for the required system instantaneous field of view, IFOV = 0.0002 rad. Indicated on the figure are nominal ranges of values for both f and d that represent reasonable values. These reasonable values are just that and do not represent hard limits.

Another factor of interest is the size in the focal plane of the image of an object at the closer ranges. If the image gets large enough it will overlap several detectors. For the chosen system constraints, even the largest object ($5m^2$) at the closest range (500km) is still small compared with the projected field of view of a single detector. This is verified in Fig. 16 where a single

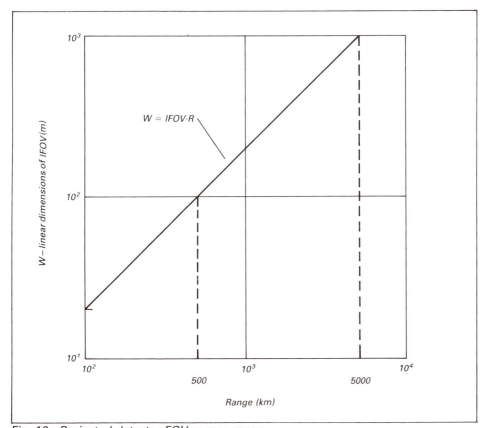

Fig. 16 Projected detector FOV versus range

Table 7   Nominal parameters – example 1

Object			
	Temperature		300K
	Emissivity		$\varepsilon = 0.8$ (all λ)
	Size		$A_s = 1$–5m^2
	Spectral band		8–14μm nominal
Path			
	Space		Transmittance $(T_a) = 1.0$
	Range		500–5000km
Optics			
	Efficiency		$T_o \approx 0.7$
	Diameter		D (to be determined)
	$f/\#$		1.5–3.0 nominal
	TFOV		175 × 1275mr
	IFOV		0.2 × 0.2mr
Detector/electronics			
	Detectivity		D^* (to be determined)
	Size		10–200μm nominal
	Visibility factor		$V \approx 1.2$ nominal
	Dwell time		$t_d = 1.3$μs (worst case)

detector views a 100 × 100m area at the closest range. Thus, energy collected by the sensor from an object will be focused to an image that is essentially a point over the entire operational range of the sensor system.

Based on the above given objectives and discussions of various parameters, Table 7 gives a list of nominal parameters. Clearly, some parameters are still to be determined.

There are no unique methods for proceeding to determine the value of certain parameters. Indeed, there are no unique solutions to a sensor design example. However, as a starting point the following techniques give valuable insight into the important criteria for choosing parameter values. The approach is in terms of determinating the effect of parameters on resolution and on signal-to-noise ratio. In fact, resolution was a primary factor in the determination of several parameters already. Next consider factors affecting the signal-to-noise ratio.

Clearly, an important factor in determining sensor performance is the optical signature of the objects to be detected. For this example the optical signature is computed as the irradiance at the sensor from the objects to be detected. Since these objects vary in size and can be at different ranges, the description is best given in graphic form. Specifically, the irradiance at the sensor due to emitted grey-body radiation is given by:

$$E = \varepsilon M_{\triangle\lambda}A_s T_a / \pi R^2 \,(\text{W/m}^2) \tag{45}$$

The total optical signature must include flux that is reflected from strong ambient sources (primarily the earth and sun). However, as a simple first cut (conservative as well), the reflected component is ignored (recall that the reflectance is ony 20%). A more detailed analysis including the reflected signature should be done only if necessary to achieve a workable design. With this justification, the emitted object irradiance at the sensor is plotted as a function of range for the two extremes of object size in Fig. 17. For the

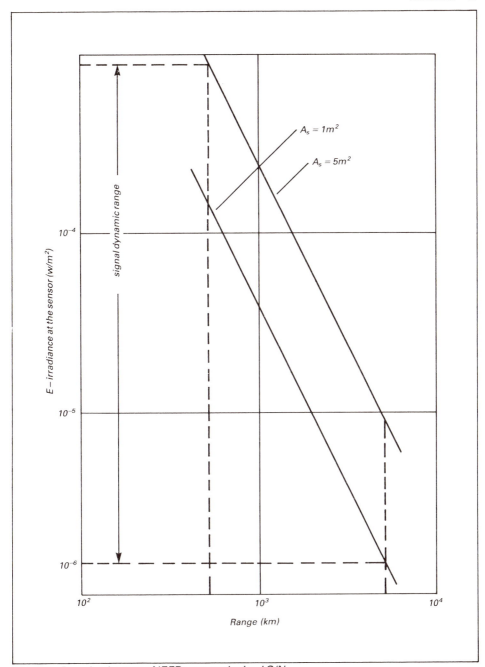

Fig. 17   Required sensor NEFD versus desired S/N

specified range values of 500–5000km, it is clear from the figure that the optical signature covers a dynamic range of approximately three orders of magnitude. As another observation, it can be said that the worse-case optical signature occurs for the smallest object at the longest range and is given by an irradiance at the sensor of $E = 1.8 \times 10^{-6}$ W/m^2.

As a next step in the sensor analysis, the requirement that the sensor is to provide object detection as opposed to recognition allows a minimum signal-to-noise ratio to be established. Although it is not presented here, a relatively simple mdoel is available for computing probability of detection and probability of false detection for a pulse detection system. From binary decision theory and an assumed Gaussian model for the noise, it can be shown that a S/N of 10 with a detection threshold S/N of 6 provides a detection probability in excess of 0.99 and a false detection probability of less than 10^{-6}. Thus, it is possible to set a minimum S/N for the sensor under worst-case conditions (e.g. smallest object, longest range, conservative optical signature) and proceed to find other parameter values to meet this minimum design goal.

Based on a minimum design goal for S/N of 10 as justified in the above paragraph, and based on knowledge of the expected optical signature as shown in Fig. 17, a useful sensor design requirement is given by the sensor noise equivalent flux density (NEFD). The relationship given by:

$$S/N = E(\text{at sensor})/\text{NEFD} \qquad (46)$$

is plotted in Fig. 18 as NEFD versus S/N. Curves are given for the worst case (smallest object, longest range) and the best case (largest object, shortest range). A design goal NEFD of 1.8×10^{-7} W/m^2 corresponds to a minimum S/N ratio of 10 under the worse-case condition.

The advantage of the above approach to sensor design is that it separates the sensor parameters from the object and path optical properties for many sensor systems. This allows a simple parametric analysis of sensor parameter interaction to try and achieve the NEFD design goal. Unfortunately, not all sensor systems allow this type of separation. For example, systems where the optical signature of the object and optical properties of the path influence achievable detector performance, the separation is not complete. For this example it has already been assumed that detector D^* is determined by background radiation from the optics. Because of this assumption and the free space optical path, this example is well suited to a separate analysis of sensor and scene.

The next step is to examine the parametric interactions affecting NEFD and to start identifying values for selected parameters. The sensor NEFD is given by:

$$\text{NEFD} = \frac{4}{\pi T_o D^2 D^*} \left(\frac{A_d}{t_d V} \right)^{1/2} \ (\text{W/m}^2) \qquad (47)$$

which is easily verified from Eqn. (41) for S/N, Eqn. (46), and from a calculation of irradiance at the sensor due to an object at range R. Irradiance at a distance R is given by:

$$E(\text{at sensor}) = \varepsilon M_{\triangle\lambda} T_a A_s / \pi R^2 \ (\text{W/m}^2) \qquad (48)$$

From this development it should be clear that the noise equivalent flux density is the equivalent irradiance on the front of the sensor that produces S/N of unity.

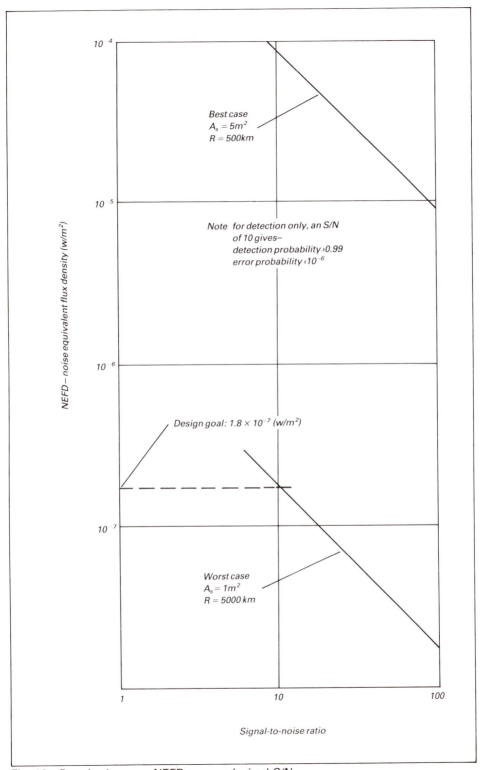

Fig. 18 Required sensor NEFD versus desired S/N

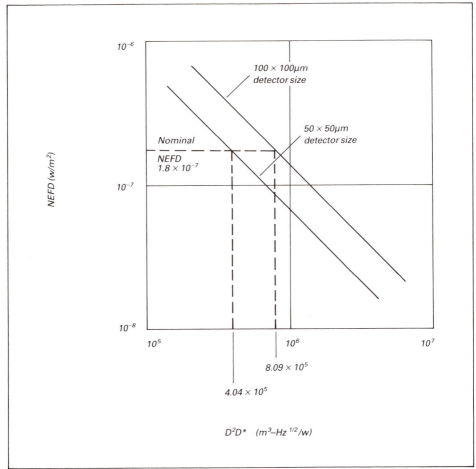

Fig. 19   NEFD versus diameter-squared detectivity product (D^2D^*)

Some of the parameters in Eqn. (47) have already been assigned values or at least been given a range of possible values. The two primary parameters still to be determined are the optics diameter D and the specific detectivity D^*. Fig. 15 gave a nominal range of detector sizes from 10 to 200µm. Within this range two values (50 and 100µm) are selected for further analysis. These values are typical of several large-scale linear arrays.

In Fig. 19 NEFD is plotted as a function of the product D^2D^* for detector sizes of 50 and 100µm. The nominal design NEFD selected previously is marked on the figure, from which a range of values from min to max are identified for the required D^2D^* product.

Next, in Fig. 20 the value of D^* is plotted versus D^2 for the two limiting cases on their product (max and min). For this example a practical limit was placed on the optics' primary diameter. Specifically, optical elements with a diameter from about 5 to 30cm are relatively easy to fabricate. As the diameter gets larger than about 30cm the difficulty and cost grow rapidly. From the identified range of possible diameters for the optics, the figure gives a range of values for specific detectivity that must be achievable.

In evaluating the feasibility of achieving the necessary values for specific detectivity, one only has to look at the performance of available detectors. D^* values in the range of 10^9–10^{11} are advertised for detectors in the 8–14μm spectral band. Realising that these values are based on a 2π steradian field of view and a 300K background, significant improvement is expected for the operating conditions in this example. The background temperature is significantly less than 300K and the field of view of any detector is simply the numerical aperture of the optical system. Using the background-limited computation for specific detectivity it can be shown that the required D^* is more than achievable.

With the above justification and delay in computing the actual D^*, the design proceeds by selecting the remaining parameter values. First, within the range of possible optics diameters, a value of 12.5cm is chosen. Clearly, quality optics are achieved at more reasonable cost if the size is not too

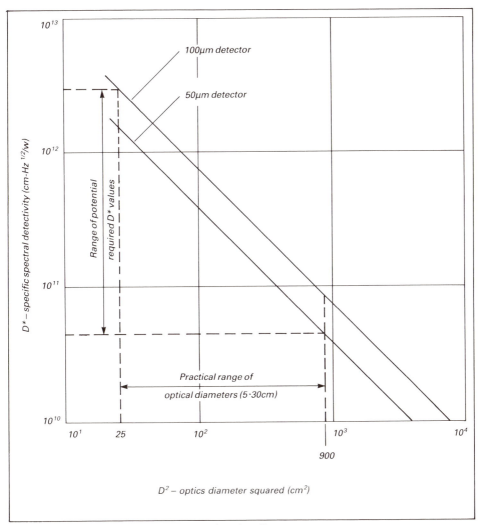

Fig. 20   Constraints on detectivity and optics diameter

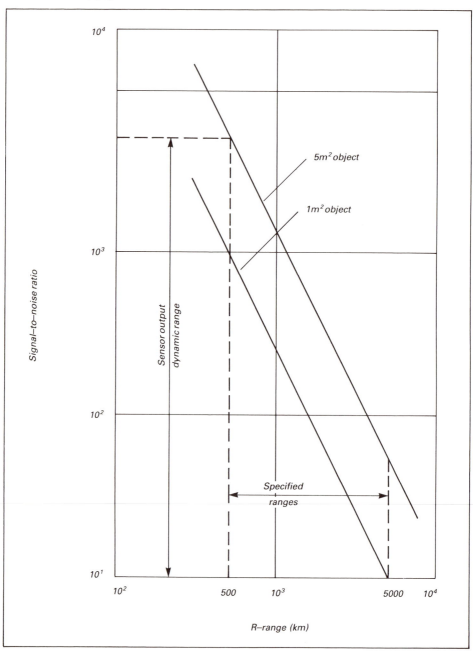

Fig. 21 S/N versus range for selected sensor design

great. Smaller diameters would work from the standpoint of *S/N*; however, the resolution requirement becomes more difficult with smaller optics. For example, if a detector size of 50μm is chosen as the smallest available at reasonable cost; then, from Fig. 15, the optics' focal length must be 0.25m or 25cm. With the chosen daimeter of 12.5cm the resulting optical system has an effective *f/#* of *f/2*, which is reasonable.

Table 8   Parameter selection – example 1

	Parameter	Value	Status
Object	Temperature	300K	Given
	Emissivity	$\varepsilon = 0.8$ (all λ)	Given
	Size	$A_s = 1\text{--}5\text{m}^2$	Given
	Spectral band	8–14μm (nominal)	Selected
Path	Space	$T_a = 1.0$ (all λ)	Assumed
	Range	$R = 500\text{--}5000\text{km}$	Given
Optics	Efficiency	$T_o \approx 0.7$	Assumed
	Diameter	$D = 12.5\text{cm}$	Selected
	f/#	2.0	Selected
	TFOV	$175 \times 175\text{mr}$	Given
	IFOV	$0.2 \times 0.2\text{mr}$	Calculated
Detector/electronics	Detectivity	$D^* = 2.5 \times 10^{11}$	Selected
	Detector type	HgCdTe	Selected
	Size	$50 \times 50\mu\text{m}$	Selected
	Visibility factor	$V = 1.2$	Assumed
	Dwell time	$t_d = 1.3\mu\text{s}$	Selected
Performance	$S/N_{min} = 10$ for smallest object at longest range		

Based in the above discussions and trade-offs, all parameters have now been selected for the system. These are listed in Table 8 where each is identified as having been given, assumed, or selected based on the analyses. Using these parameter values, Fig. 21 gives the sensor S/N as a function of range from the smallest and largest objects. From the figure it is clear that the S/N goes from a worst case of 10 to approximately 50,000 over the required operating range. Obviously, another input to detector selection is that it must be able to handle a dynamic range of 5000 to 1 for this application.

Some final comments need to be made about the above sensor design example. First, the selected parameters in Table 8 represent only one design that works. The choice of this particular design versus another workable design makes no difference unless additional constraints are placed on the system. Secondly, the approach used in analysing the system and the order in which parameters were selected is not unique nor necessarily the best. Different insights are achieved sometimes by changing the approach to a problem. Finally, the variability of analysis methods and design solutions serves as a challenge and almost demands creativity. It is the purpose of this example and the next to provide awareness of the challenges involved in sensor analysis, to give some useful approaches to the analysis, and to stimulate creativity in the solution.

Performance evaluation: imaging sensor

The next example is different in several aspects from the previous example. First it requires an imaging sensor and object identification so that a higher S/N is necessary. However, the application is a close-range controlled scenario such as might be found in an industrial automation environment.

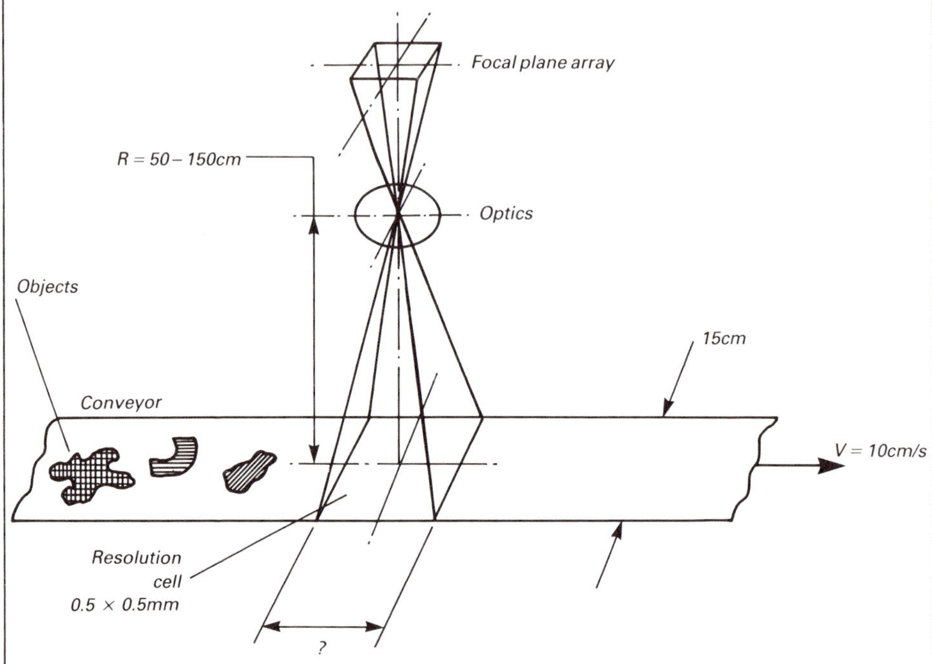

Fig. 22   Sensor system example 2 geometry

Specifically, the sensor must provide high-resolution, high-quality images of objects moving on a conveyor belt. A system diagram of the example is given in Fig. 22 with some of the parameters identified. Because the system is to operate in a controlled environment with the possibility of adequate ambient lighting, it is likely that a sensor operating in the visible spectrum is the best choice. Further supporting this argument is the fact that high-sensitivity imaging sensors are most readily available for the visible spectrum, whether they be imaging tubes such as the vidicon or two-dimensional focal plane array sensors using silicone-based detectors.

Because of the availability of imaging sensors as packaged units for the visible spectrum, this example will proceed in a very different manner than the previous example. However, the same performance measures are still the driving considerations. For example, the resolution requirement and sensor/scene dynamics strongly impact the number of detectors needed and the methodology for selecting and processing image frames.

In designing the system, there is an advantage to choosing frame rate and timing requirements to be compatible with commercially available equipment. Specifically, if the sensor timing is compatible with commercial television then an inexpensive display can be used to view the sensor output. If the timing is radically different then the system must provide for specialised display electronics. The point is that compatibility with existing equipment can be a very strong overall system constraint on the new sensor system. Clearly this constraint is acceptable only if it is beneficial to the overall system performance.

Based on the above initial arguments, a first-cut design choice is to use a two-dimensional focal plane array imaging sensor operating in the visible spectrum. The next step in the analysis is to determine the size array necessary to meet the resolution requirements and relate this to the sensor/scene dynamics.

A frame is defined as a complete two-dimensional image obtained by sampling each detector in the array only once. Since detectors are typically sampled sequentially in time, the total time it takes to sample every detector once is called the frame time t_f. The frame rate is then the reciprocal of the frame time. For standard television systems the frame rate is 30/s so that the frame time is approximately 33ms. Thus, a choice of 33ms for the frame time will make the system compatible with inexpensive television displays. Based on the chosen frame time and using the given velocity of the conveyor belt it is clear that the belt moves in one frame time a distance given by:

$$d = t_f v = 3.3mm \tag{49}$$

The system is required to identify objects within its field of view. An approach that is compatible with available image processing equipment is to capture a single image frame and process it to identify objects. That is, a single complete image scan is stored in memory for processing. The amount of time allowed for the processing before another image must be 'grabbed' is determined by the system dynamics and the total field of view of the sensor.

Commercially available cameras with focal plane arrays typically have a rectangular array of detectors. The aspect ratio (number of detectors in one direction as compared with the other direction) is typically 1:1, 2:3, or 3:4. Furthermore, for this system the field of view across the belt should be at least as wide as the belt or 15cm. Based on this constraint and the available aspect ratios, the field of view along the belt can have values from 10 to 22.5cm. The choice depends on several criteria that will generally have varying relative importance for different applications. A choice of 20cm is made for the example system based on a selected set of criteria as explained below.

First and most importantly it is essential that all objects be identified. The identification is accomplished by freezing an image every t_{ff} seconds and processing it using appropriate algorithms. A corollary to this requirement is that any given object must never move a distance greater than the field of view along the belt between freeze frames. If an object must be completely within the field of view to be identified then a limiting case for the relationship between t_{ff}, belt velocity, and field of view is as shown in Fig. 23. An object moves a distance:

$$d_{ff} = t_{ff} v \tag{50}$$

in one freeze-frame time. Given the constraint that the largest object must be completely within either frame 1 or frame 2 requires that for the case

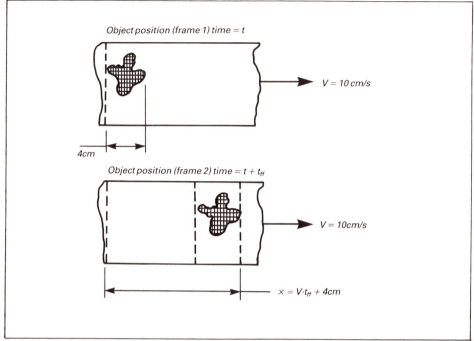

Fig. 23   Total FOV dependence on processing time

shown, the object will be completely within both frames. The required total field of view along the belt is then given by:

$$X = d_{\text{ff}} = + d_{\text{object}} \tag{51}$$

where the largest object size is 4 cm. Using this value and the belt velocity gives:

$$X = 10t_{\text{ff}} + 4 \text{ (cm)} \tag{52}$$

Clearly, the time available for processing an image (t_{ff}) is directly proportional to the field of view along the belt, X. To allow a sufficient amount of time for image processing, it is then preferable to choose one of the larger values for X. Although it is not the largest, a value of $X = 20$cm is chosen because it is compatible with a standard television display aspect ratio of four horizontal to three vertical. With this value of X, the freeze-frame or processing time is then:

$$t_{\text{ff}} = \frac{20 - 4}{10} = 1.6\text{s} \tag{53}$$

There are clearly situations where an object will appear in the left side of one frame and also in the right side of the next frame. This redundant imaging will have to be handled by the processor.

One final note with respect to the system dynamics gives the amount of blurring of an object. For an object with a maximum size of 4cm, the system provides a complete scan of the object in $(4/20)t_f = 6.6$ms (for across-belt scanning) or in $(4/15)t_f = 8.8$ms (for along-belt scanning). The resultant image has a maximum blur due to movement of either 0.66 or 0.88mm depending on the scan direction. This blur is only slightly larger than the resolution of 0.5mm so that relatively little imaging degradation is expected.

Since total field of view and instantaneous field of view are now specified in terms of dimensions on the belt, it is possible to determine the number of detectors required for the focal plane array. The result is complicated by the stated requirement to have the sensor operate at distances from 0.5 to 1.5m from the belt. The instantaneous field of view given in angular measurement is dependent on the distance as is the total angular field of view. In order to meet the stated resolution of 0.5mm on the object at a distance of 1.5m, the angular resolution must be no more than 0.5mm/1.5m = 0.33mrad. Additionally, in order to cover the full 15×20cm^2 total area, the sensor must have a total angular field of view at 0.5m distance of 15cm/0.5m $\times$ 20cm/0.5m = 300×400mrad. And finally, if the system is to meet both these requirements without variable focus optics the number of detectors required is:

$$\frac{\text{TFOV}_{max}}{\text{IFOV}_{min}} = \frac{300 \times 400 \text{ mrad}}{0.33 \times 0.33 \text{ mrad}} = 900 \times 1200 \qquad (54)$$

This large number of detectors is impractical and probably will be for the foreseeable future, so a compromise is indicated. The most reasonable compromise is to choose one nominal operating distance. At a fixed distance, the number of detectors required becomes 300×400 as indicated in Fig. 24. Interestingly enough, the number of detectors is the same for any

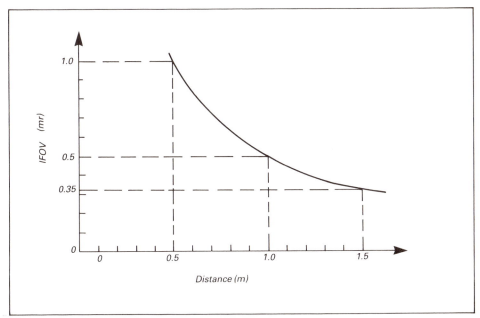

Fig. 24   Resolution limits

Table 9   Nominal parameters – example 2

Object	Three sizes	$\approx$ three, 3 and 4cm
	Shape	3cm objects similar
		4cm object elongated
	Spectral properties	3cm objects, $\rho = 0.6$ and 0.4
		4cm object, $\rho = 0.6$
	Spectral band	Visible
Path	Air	Close-range, controlled
	Range	50–150cm
Optics	Efficiency	$T_o \approx 0.7$
	Diameter	D (to be determined)
	$f/\#$	1.5–3.0 nominal
	TFOV	15×20cm at object
	IFOV	0.5×0.5mm at object
Detector/electronics	Detectivity	D^* (to be determined)
	Update time	$t_{ff} = 1.6$s
	Frame time	$t_f = 1/30$s
	Dwell time	$t_d = 0.2777\mu$s
	Array size	300×400 elements
	Integration time	$t_i = 1/30$s

fixed distance within the specified range of distances. The thing that must change from one distance to another for a given fixed-size focal plane array is, of course, the optics' focal length.

Based on the above discussions and the system objectives stated in Table 6, a revised set of nominal parameters is given in Table 9 for this example. As another check on the practicality of parameters selected so far, a 50mm $f/2.0$ lens has a focal length of 100mm. This lens used with a focal plane array that has 50μm detector spacing does provide a spatial resolution of 0.5mrad which is the specified resolution for a distance of 1m. All these numbers are compatible and nominally practical; therefore, they will be used in the first-cut design.

The next major step in the system analysis is to determine the signal-to-noise ratio expected. This problem is much simpler for this example than for the first example presented because the environment is controlled and sufficient ambient illumination can be made available to provide the desired operating range. Best results are obtained if the system is designed to operate near the middle of the sensor dynamic range. Most manufacturers provide performance curves that give the output signal-to-noise ratio versus faceplate illumination for the sensor. These curves are typically given in terms of photometric units for a given source (such as a tungsten lamp at a given temperature) and for a specific lens. If the conditions under which the system is to be used are radically different then compensation may be necessary. Fortunately, most available camera/lens combinations are designed to operate with high *S/N* under 'ordinary' illumination levels.

Several solid-state camera systems are available with the desired number of detectors and with high signal-to-noise ratios under typical ambient lighting conditions. A nominal system will require illumination levels of about 2–5 lm/m^2 for output signal-to-noise ratios of about 80–200. The signal processor which must identify the objects will probably require high-quality images (e.g. S/N of at least 100).

As an example of the level of ambient illumination needed to provide S/N of 100, consider the following. An ambient source of E_a (lm/m^2) is incident upon the scene to be imaged. The illuminance at the faceplate is then given by:

$$E_s = E_a A_{FOV} \rho_{min} / \pi R^2 \ (lm/m^2) \tag{55}$$

or in terms of the required ambient illumination:

$$E_a \ (required) = E_s \pi R^2 / A_{FOV} \rho_{min} \ (lm/m^2) \tag{56}$$

Using the scene area in the total field of view (15×20cm) and the nominal sensor sensitivity of 2.5 lm/m^2 to get $S/N = 100$ then requires an ambient illumination of:

$$E_a = \frac{(2.5)(\pi)(1)^2}{(0.15)(0.20)(0.4)} = 654 \ lm/m^2 \tag{57}$$

This level of ambient illumination is easily achieved with simple lighting arrangements. The remainder of the problem in this example deals with the kinds of image processing methods necessary to accomplish the object identification in the allowed time of 1.6s/image. The details of the image processing are not intended to be part of this presentation.

As a final summary to the examples presented in this article, it is emphasised again that no single design is correct for any given sensor application. Rather, there are many possible solutions and the best choice is usually dependent on the specific constraints and objectives for the application. Additional experience in the design and analysis of electro-optical sensors can be obtained through practice.

Appendix

Conversion between radiometric and photometric units

Curves for spectral luminous efficiency give the relative spectral response for an average eye. Plots of these spectral curves are given earlier. Two spectral functions are defined: $V(\lambda)$ is the spectral luminous efficiency for photopic vision and $V'(\lambda)$ is the spectral luminous efficiency for scotopic vision. These experimentally derived curves have relative maxima at $\lambda = 0.555\mu m$ for photopic vision and $\lambda = 0.51\mu m$ for scotopic vision.

Spectral luminous efficacy provides a conversion between radiometric flux (in watts) and luminous flux (in lumens). Efficacy is defined for both photopic and scotopic vision as:

$$K(\lambda) = K_{max} V(\lambda) \ (lm/W) \tag{A1}$$
$$K'(\lambda) = K_{max} V'(\lambda) \ (lm/W)$$

where the maxima for efficacy occur at the same wavelengths as the maxima for efficiency. The values for K_{max} and K'_{max} are determined from the definitions for lumen and the standard candela.

The candela is the basic unit of luminous intensity (lm/sr) and is defined as the luminous intensity from a $1/60\text{cm}^2$ black-body at the solidification temperature of platinum (which is 2048K).

To convert any radiometric source flux to photometric units is accomplished with the use of the spectral luminous efficacy as:

$$\phi_p = \int_0^\infty \phi_p(\lambda)\, d\lambda = \int_0^\infty K(\lambda) \cdot \phi_r(\lambda)\, d\lambda \text{ (photopic)}$$

$$= \int_0^\infty K'(\lambda) \cdot \phi_r(\lambda)\, d\lambda \text{ (scotopic)}$$

(A2)

where the subscripts p and r denote photometric and radiometric quantities, respectively. Using Eqns. (A1) and (A2) gives:

$$\phi_p = \int_0^\infty K_{\max} V(\lambda) \cdot \phi_r(\lambda)\, d\lambda \text{ (photopic)}$$

$$= \int_0^\infty K_{\max} V'(\lambda) \cdot \phi_r(\lambda)\, d\lambda \text{ (scotopic)}$$

(A3)

The result in Eqn. (A3) is valid when $\phi_r(\lambda)$ is replaced by an appropriate spectral radiometric quantity if $\phi_p(\lambda)$ is replaced by the equivalent photometric quantity.

To find the value of $K_{\max}$, using the definition for the standard candela gives Eqn. (A3) in terms of intensity as:

$$I_p = 1(\text{lm/sr}) = \int_0^\infty K_{\max} V(\lambda) \frac{(M(\lambda,2048K)/\pi)(1/60)\, d\lambda}{I_r\,(\text{W/sr})}$$

(A4)

with a similar equation for scotopic vision. Solving for the maximum values of efficacy gives:

$$K_{\max} = \frac{60\pi}{\int_0^\infty V(\lambda)M(\lambda,2048K)\, d\lambda}$$

$$K_{\max} = \frac{60\pi}{\int_0^\infty V^l(\lambda)M(\lambda,2048K)\, d\lambda}$$

(A5)

The integrals in Eqn. (A5) may be integrated numerically to give the values $K_{\max} = 673$ (lm/W) and $K_{\max} = 1725$ (lm/W).

From the above results is derived a measure of the overall efficacy of any given source. The total luminous efficacy for a source with spectral radiant flux $\phi_r(\lambda)$ is:

$$K = \frac{\int_0^\infty \phi_p(\lambda)\, d\lambda}{\int_0^\infty \phi_r(\mu)\, d\lambda} = \frac{\int_0^\infty V(\lambda) \cdot \phi_r(\lambda)\, d\lambda}{\int_0^\infty \phi_r(\lambda)\, d\lambda}$$

(A6)

Values of K are tabulated for some common sources such as tungsten lamps and others.

References

[1] Stimson, A. 1974. *Photometry and Radiometry for Engineers.* Wiley, New York.

[2] *RCA Electro-Optics Handbook,* EOH-11. RCA Corporation, Harrison, NJ, USA, 1974.

[3] Wolfe, W. L. 1965. *Handbook of Military Infrared Technology.* Office of Naval Research, Washington, DC, USA.

[4] Petritz, R. L. 1959. Fundamentals of infrared detectors. *Proc. IEEE,* 47 (September): 1458–1466.

[5] Swift, I. H. 1961. Performance of background-limited system for space use. *Infrafed Phys.,* 2: 19–30.

[6] Genoud, R. H. 1959. Infrared search-system range performance. *Proc. IEEE,* 47 (September): 1581–1585.

2

Special Vision Sensors

The availability of commercial vision sensors specifically designed for robotic applications are long overdue. Some developments in vision sensors specifically designed for robotic applications are covered.

A LOW-RESOLUTION VISION SENSOR

D. G. Whitehead, I. Mitchell and P. V. Mellor
University of Hull, UK

A low-resolution camera system based on a 64K dynamic RAM chip is described. The camera is capable of providing a picture resolution of up to 256 × 128 pixels. A fast microprocessor is used to collect the picture data and to provide a preprocessing facility and a flexible means of access to the data.

A camera system of low resolution (up to 256 × 128 pixels) which is eminently suitable for such purposes as component inspection and identification is described. The use of a low-cost dynamic RAM chip as a light sensor provides a method of imaging not really obtainable by any other means. Area CCD cameras (as distinct from line-scan arrays), such as the Sony XC-37 and Fairchild CCD3000, though providing a greater resolution, are an order of magnitude greater in cost: ranging from £700 to some £2000. These types of camera are not strictly comparable with a dynamic RAM camera as they give grey-scale imaging directly and are designed to be compatible with television displays. As such, imaging data is not as easily obtainable from them in a form suitable for input to a microcomputer system. Also, the larger resolution of the CCD camera is not always necessary for parts inspection and identification. In such cases, a RAM camera system, costing only around £100 to build, can be a more appropriate instrument.

Details are given in this paper of the circuits used to interface to the camera. The visual data is obtained in a form convenient for capture and subsequent processing by a microcomputer system.

The camera

Construction

The camera is a simple structure using as the light-sensitive element a dynamic RAM chip specifically constructed for optical work: the silicon chip is housed within a ceramic package with a built-in window to allow light to be focused onto the active part of the silicon. This device, the IS32, is a development of the μT4264 64K DRAM (Micron Technology Inc.). A small projector-type lens (focal length 9mm) is used to focus the image (Fig. 1).

Fig. 1   The camera unit

The lens is threaded, allowing focal adjustment. This is not critical, the effect of exposure time and focus being rather similar in that memory cells on the fringe of images may or not be affected depending on either the sharpness of the image or the sensitivity of the array. In practice we have found that a combination of 'coarse' focal length and a lens stop can be chosen to suit the application and 'fine tuning' effected by means of the exposure time. It should be noted that, unlike, say, a Vidicon tube, overexposure has no detrimental effect on the RAM chip.

The camera body is constructed from black 'nylatron' plastic and connection to the drive electronics is via 16-way ribbon cable. The overall size of the camera is 2.5 × 2 × 3cm.

The light-sensitive array

The IS32 is a 64 × 1 bit dynamic RAM chip laid out as two 32K bit arrays, physically separated by the sense amplifiers. As a consequence of this, only one-half of the available RAM can conveniently be used for the picture array.

Data are held in a DRAM by charge storage. In 'normal' use, this charge would gradually leak away due to thermal effects and the information would be lost were it not that the charge is replenished by 'refresh' circuitry built into the chip which replaces the charge automatically upon row selection during the reading process. It is therefore necessary to ensure that the entire array of a DRAM is read at regular intervals so that refreshing can take place. In microcomputer systems dynamic RAM refresh takes place automatically and is transparent to the user.

As well as thermal effects, photoelectric current will discharge the data cells and it is this photoelectric effect that is exploited to provide a light-sensitive array. By precharging the memory cells and reading their state at regular intervals a representation of light incidence on the array can be obtained in terms of 'ones' and 'noughts'[1]. The image obtained by this process is binary in nature, the amplifiers within the DRAM interpreting any level of charge as one of the two logic levels. It is possible to produce a grey-scale image but this must be done by taking repeated 'snap shots' at different exposure times and then processing the resulting set of images into a single composite picture.

Our initial experiments on the use of commercially available DRAM devices as low-resolution camera elements centred upon the 4116 DRAM. Earlier versions of this device were housed in a ceramic package with a metal lid which could be removed (carefully) and replaced by a thin piece of glass[2]. A disadvantage of the 4116 concerns its topology – the storage cells are not arranged in straight rows, but in a diagonal manner. This gives a castellated edge effect to images. Note that a commercial camera (type 511 Image Digitizer, Periphicon Inc.) uses a DRAM chip similar to the now obsolete MK4006 DRAM (Mostek Inc.). This chip does not have on-board refresh circuitry (an advantage for imaging work) but the array size restricts resolution to 32×32 pixels. The IS32 by comparison has its cells organised in straight rows and is therefore more suitable for camera work. It should be noted that the separation of each cell on the IS32 is not equal: the row/column bit spacing is $21.5\mu m/8.5\mu m$. This means that images are distorted when viewed with a direct mapping from the DRAM to a VDU screen. For computer analysis, this is unimportant and in some instances can be used to advantage. For example, only circular objects will present a constant area (equal number of pixels) when presented to the camera and then rotated.

Exposure time

In use, a 'read-modify-write' cycle is employed: addressed cells are charged by a write cycle immediately after being read. Note that the internal arrangement of the cells demands that for half of the array logical 'one' corresponds to a charged cell and for the other half logical 'nought' indicates the presence of charge. Connecting the most significant address bit to D_{in} (Fig. 3) takes care of this inversion. The read cycle time relates to the 'exposure' time directly. The automatic refresh that takes place during each row selection has the effect of 'setting' the charge level in the cells to either the 'one' or the 'nought' level. Thus the effective exposure time is the time between each row selection. For a continuously running clock the exposure time is given by

$$t_{exp} = (N_R - 1) \cdot N_c \cdot t$$

where N_R is the number of rows, N_C is the number of columns and t is the system clock period. In the present design, t can be varied between one and six microseconds, giving a wide range of exposure times.

Hardware

To obtain correct exposure times under normal lighting levels it was found that a clock rate of approximately 0.8MHz was necessary for the DRAM. At this rate, transfer of data from the IS32 to a RAM store presents several problems. Either discrete logic must be used to manipulate the data or a fast microprocessor is required. The latter approach was chosen as it seemed to offer a greater degree of flexibility for subsequent work.

A block diagram of the microprocessor system is shown in Fig. 2 and the IS32 drive circuitry is shown in Fig. 3. The camera is controlled by an Intel 8031 microprocessor. This device can operate at clock speeds of up to 12MHz. The camera clock is derived from the processor clock and through the 74LS393 counters generates the memory addresses for the IS32. The exclusive-OR gates are required to 'unscramble' the IS32 addresses since the physical positions of the cells on the chip do not correspond to their logical positions. Serial data from the DRAM, are converted into 8-bit parallel data bytes by the 74LS164 shift register, clocked into the 74LS374 8-bit latch and then, through the 8-bit driver, 74LS241, transmitted via ribbon cable to the processor.

By using a clock signal for the IS32 which is derived from the 8031 clock circuits, a software–hardware synchronous system is created. A 'handshake' is then not necessary for the transfer of data to the 8031 from the camera, a

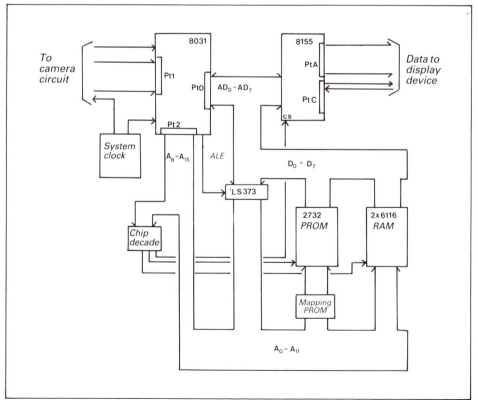

Fig. 2   The 8031 microprocessor control system

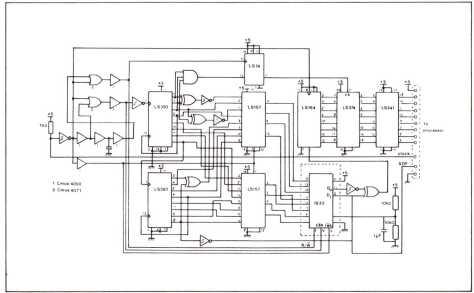

Fig. 3   The camera drive electronics

'start-of-frame' (SOF) being all that is required. The 8031 program shown in Table 1 is timed to synchronise with the data byte rate from the shift register. This program stores an IS32 picture byte into a given location in the fast 6116 RAM used as the picture store. By modifying the address decode to the PROM containing this short program it is possible to force the 8031 to repeat this code 4K − 1 times, thus enabling a complete picture frame to be stored as 8-bit bytes in successive locations in the RAM. After the 4K iterations of this program section the 8031 'falls through' and accesses the next section. Fig. 4 gives details of the PROM decode.

The SOF signal ensures correct synchronisation of the data capture sequence. Since that data is present for at least eight microseconds, sufficient time is available to branch into the capture sequence following the SOF signal. Note that this SOF signal, derived from the IS32 address, provides a position-going edge at the commencement of addressing the first 32K block of the IS32 DRAM and a negative-going edge when the second block is addressed. By altering the polarity of the edge detector, either half of the IS32 may be selected. When a complete frame from the IS32 (32K bits) has been captured, the 8031 can be used to process the data. In our present system, only limited modification is made before outputting the data to a

Table 1   Data capture program for the 8031

READ:	NOP	Waste some time
	MOV A, P1	Read in data byte from IS32
	MOV X DPTR, A	Store video data in RAM
	INC DPTR	Point to next free location
	ORL ACC, $OOH	Waste more time here
READ LOC:		Entry point to recurrent 8 byte image stream

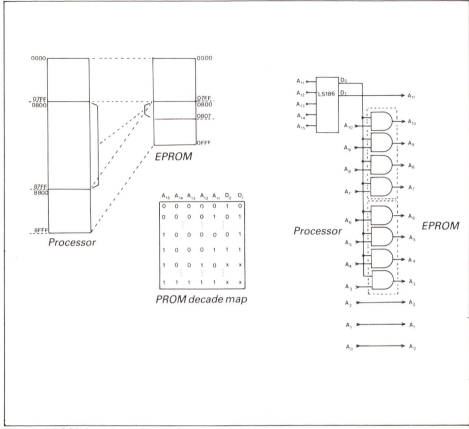

Fig. 4   PROM decode and mapping

display system (Motorola Exorset). The processing carried out on the pictures shown (Fig. 5) was simply to reduce the amount of data by masking out the odd-addressed bits and then duplicating the even ones. This maintains a 4K byte frame but reduces considerably the picture noise. This noise appears to be caused by a 'nearest neighbour' disturbance effect or simply the result of light fringing. The effect is to produce a chequered pattern of pixels at light/dark boundaries. Selecting alternate pixels removes this but with the penalty of a reduction in resolution to 64×128 pixels. The pictures show that even with this small amount of processing excellent results can be obtained.

The Motorola Exorset accepts the data as 8-bit bytes from the 8031 through its standard 6821 port interface and is used simply as a convenient medium for display and experimental processing of the images.

Concluding remarks

A simple low-cost camera system for use in inspection and parts identification work has been described. Good-quality picture resolution of 64×128 pixels can readily be obtained from a 128×256 pixel base. Details of an image processor for the video data have been given but the camera is

Fig. 5 (a) A self-tapping screw; (b) single- and double-sided pcb pins; (c) shake-proof and circular washers. Note the distortion caused by the unequal aspect ratio of the pixel spacing

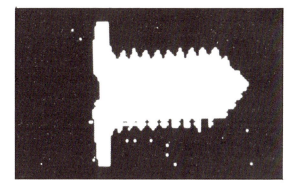

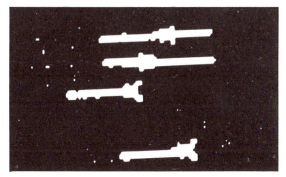

suitable for use with most computer systems having an interface capable of accepting serial data at a 1MHz bit rate. The use of the Exorset has been found to be most convenient for our image-processing development work – in particular recognition of edges and beacons and the calculation of centroids. Such facilities will eventually be transferred to the 8031, which will then function as an 'intelligent' slave processor in the multiprocessor system[3] under development within the Department of Electronic Engineering for use in the investigation of robot assembly tasks.

Acknowledgements

This work has been funded by the Science and Engineering Research Council under contract number GR/B/4586.3 and GR/B/5940.2.

References

[1] Russell, R. A. 1983. Computer vision system for applications in robotics education. *Microprocessor and Microsystems*, 7 (September): 320-32

[2] Hodgson, R. M. 1982. Dynamic RAM chips as optical array sensors. Internal Report, Department of Electronic Engineering, University of Hull.

[3] Mitchell, I., Whitehead, D. G. and Pugh, A. 1983. A multi-processor system for sensory robotic assembly. *Sensor Review*, April 1983: 94-6.

AN INTEGRATED VISION/RANGE SENSOR

J. E. Orrock, J. H. Garfunkel and B. A. Owen
Honeywell, Inc., USA

A programme is described whose objective was the development and demonstration of an integrated vision/range sensor applicable for robotic vision and inspection applications. The sensor developed combines a 2-D vision array and sparse (single-point) range sensing in one package approximately $5 \times 7 \times 10$cm in size. The sensor operates over a measurement range of 10–100cm and has a range resolution of better than ± 0.05cm at 10cm and ± 3cm at 100cm. The vision and range functions operate with a fixed-focus lens, eliminating the need for a focusing mechanism. The lightweight sensor can be arm mounted to provide part identification, acquisition and mating information, or world mounted as a verification and monitoring sensor. A review of the state-of-the-art in range sensing led to the selection of a solid-state device, used for focusing SLR cameras, to provide the ranging function for the sensor. A standard CID array is used for vision. The sensor optical design is discussed along with the range algorithms developed and the sensor performance characteristics.

With present technology, a great majority of industrial tasks, including assembly, are beyond current robots' capabilities. Several key developments in robotics technology are needed to make the transition from today's less flexible robots to the intelligent robots required in the future's less constrained environments. Among the foremost of these developments is the evolution of sophisticated robot sensing devices. These devices include noncontact sensors, such as those for vision, range and proximity, and contact sensors such as those for force and touch.

The vision/range sensor design described here is based on the integration of an off-the-shelf 2-D visible imaging array with a solid-state ranging device. This device, called the through-the-camera-lens (TCL) sensor, was developed for automatic focusing in SLR cameras. In SLR cameras, a mirror which normally reflects light coming through the lens to the photographer's eye is replaced with a partially transmitting mirror. This mirror reflects some of the light to the photographer's eye and allows some to pass through to become incident on the TCL sensor. In this way, the TCL looks 'through the camera lens' at the scene (see Fig. 1).

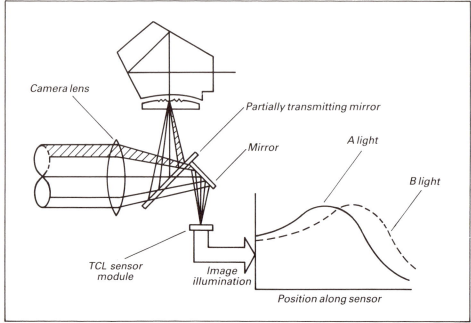

Fig. 1   TCL operation in an SLR camera

The TCL outputs a pair of signatures which can be compared for camera focusing, or in this application, ranging. The signatures are generated by 24 pairs of CCD detectors on the TCL sensor. One detector in each pair receives light coming through one sector of the camera lens (A light) and the other light coming through another (B light). Together, all A detectors register one signature (the variation of scene luminance in the portion of the image which falls on the row of microlenses) and the B detectors another. These signatures are identical when the lens is in focus, but one is displaced from the other when the lens is out of focus. The amount of displacement is proportional to the focus error (see Fig. 2).

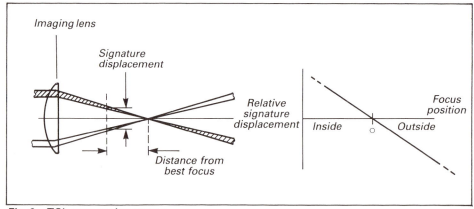

Fig. 2   TCL sensor signatures

The imaging array used in the vision/range sensor was that from a General Electric TN2200 camera. The array has 128×128 pixels. This resolution was chosen to facilitate visual verification and identification tasks.

The vision/range sensor optical design is based on a fixed-focus concept, i.e. both the vision and range sensors operate with the objective lenses set at fixed focal points. Since the sensor is developed for robot-arm-mounted applications, eliminating the need for a focusing mechanism is desirable to allow minimising sensor size, weight and complexity. The specified measurement range is 10–100cm, which is appropriate for an assembly robot work volume. The design constraint, then, is that both the vision and range functions must operate over this range in a fixed-focus mode.

The TCL requires an f/2.8 or faster lens. A 45mm focal length, 20mm achromat lens was chosen for this design. The optically active part of the TCL consists of 24 pairs of detectors, with a microlens associated with each detector pair. Each microlens is approximately 0.007in. in diameter on a 0.008in. centre. The overall size, therefore, is approximately 0.007×0.190in. In use, the TCL system is focused at about 30cm.

An analysis of the requirements for the vision lens showed that this lens must be quite different in both focal length and aperture from the range lens. The focal length of the objective lens is determined by the detector array size and the required field of view. The detector is an array of 128×128 square elements uniformly distributed within an area 0.2304in.^{2}. The requirement for the field of view is an area 12in.^{2} square at 1m object distance. For this design, a 17.5mm focal length f/6.0 air-spaced triplet lens was chosen. The vision lens focal point selection is very application dependent; for the analyses described later in this paper, a focal point of approximately 40cm was used.

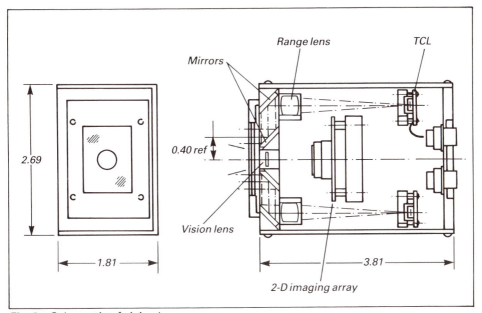

Fig. 3   Schematic of vision/range sensor

A scale drawing of the sensor package is shown in Fig. 3. The package size is 1.8×2.7×3.8in. The basic structure is a heavy-walled aluminium 'U'. Lenses and mirrors are mounted on the front arm of the 'U', the supports for the TCL and vision array are mounted on the base, and electrical connectors on the back arm of the 'U'. Another 'U'-shaped piece of thin aluminium forms the closure for the assembly. Wires are soldered directly to the TCL chip, which is glued to a small circuit board. The board is spring mounted to its support, providing the necessary adjustments for optical alignment. The wires are run in a loop – for strain relief – to one of the two connectors at the rear of the assembly.

The vision array chip is mounted on a 1.5×1.5in. circuit board. This board is rigidly mounted to its support. Wires are run from the circuit board to the second connector on the rear of the camera.

The front of the 'U' is a thick block of aluminium which is machined so that two flat, front-surface mirrors can be directly glued to the metal. The centre of this piece is drilled and tapped so that the vision lens, which comes with a threaded case, can be screwed directly into the block. The range lens and its associated aperture are mounted on the rear surface of this block. For this design, the vision axis and range axes are parallel and offset by approximately 1cm. A glass window, which protects the optical components and seals the case, is mounted on the front of the block. In this way, all of the optical components are mounted to one block. Note that the sensor package is designed to accommodate two TCL chips but to date only one TCL has been installed for analysis.

Substantial remoting of electronics allowed minimising the package size by including only the sensing arrays in the package. The TCL was remoted 2.8m from its electronics and the vision array was remoted 1.5m.

Sensor performance

As discussed previously, the integrated vision/range sensor combined off-the-shelf vision sensing technology with a potential range sensing technology, i.e. the TCL sensor. Since the performance of the vision sensing technology is comparatively well understood, the performance evaluation on this program was focused on the range sensing function of the sensor. The ranging performance specification called for measuring range from 10 to 100cm with an accuracy of ±0.031cm at 10cm and ±5cm at 100cm.

Range sensing performance was evaluated in two different environments: statically, on an optical bench, and dynamically, on a robot.

The following sections discuss the ranging algorithms developed to measure range based on the TCL sensor output, and the methods and results of the static and dynamic performance evaluations.

Fixed-focus ranging

Recall that when the TCL scene is in focus the A and B signatures are identical; when the scene is out of focus the signatures are displaced. This is illustrated in Fig. 4.

With the TCL in focus at a fixed point, the displacement or shift (measured in pixels) between the A and B signatures is a function of the

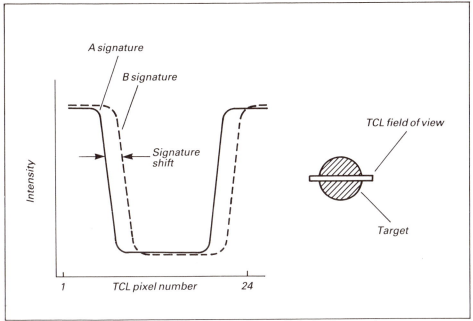

Fig. 4   TCL A and B signatures for a dark target on a light background

range to the target, where the signature shift is zero at the focal point, negative at points inside of focus and positive outside of focus. Therefore, the fundamental requirement of a fixed-focus algorithm is the ability to measure the signature shift at a given point. Range is then determined from this measured shift by using an empirically generated look-up table of range versus signature shift.

During this program several algorithms were considered for signature shift calculation, including cross-correlation techniques and adaptations of lens focusing algorithms. To date, the best performance has been obtained with an enhanced lens focusing algorithm. The algorithm is based on a parameter F_F which is a function of the intensity curve slope and the area between the A and B intensity curves (signatures). H_F is zero at the point where the A and B curves are identical, and plus or minus when the curves are shifted. H_F, then, gives a measure of the shift between the A and B signatures.

Static evaluation

Initial sensor performance evaluation was conducted statically on an optical bench with the TCL output interfaced to a microprocessor. Software was written to facilitate obtaining the TCL signatures and performing the H_F calculation. Preliminary analysis included evaluation of the H_F calculation repeatability throughout the measurement range of 10–100cm. Subsequently, H_F versus range tables were generated and implemented as look-up tables in the software to allow performing range accuracy evaluations. Fig. 5 shows the three-sigma range error versus range for the sensor. The specification is shown for reference. The range resolution obtained was approximately ±0.03cm at 10cm and ±3cm at 100cm.

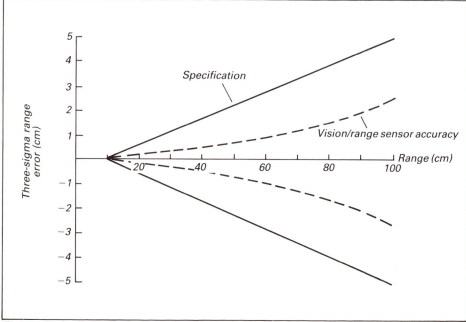

Fig. 5   Vision/range sensor static range accuracy

Dynamic evaluation

The objective of the dynamic sensor performance evaluation was to determine whether the fixed-focus ranging algorithm evaluated statically was still applicable in a dynamic (robot-mounted) environment. The requirements for the vision/range sensor in a robotic environment are to use integrated vision and range sensing for target identification and acquisition. The issues of primary concern were:

- The sensor would be robot mounted. The potential problems are lighting variations and mechanical and electrical noise.
- The sensor must interface with three-dimensional targets. The potential problems are shadowing and target complexities.
- Range and vision sensing must be integrated. The potential problem is that the TCL must handle positioning variations on the target due to vision inaccuracy.

In order to perform the dynamic performance evaluation the system shown in Fig. 6 was configured. The vision/range sensor was mounted on a PUMA 560 robot as shown. The robot controller contained the PUMA operating system and language, VAL. The TCL output was connected to a microprocessor for range processing. The vision output was connected to a Machine Intelligence Corporation (MIC) VS-100 vision processor. The VS-100 is a binary vision system which can be trained to identify and locate selected targets. The VS-100 was interfaced to VAL through a standard interface on the controller. The range processor was interfaced to VAL through the parallel I/O ports. Various targets were used for analysis. In

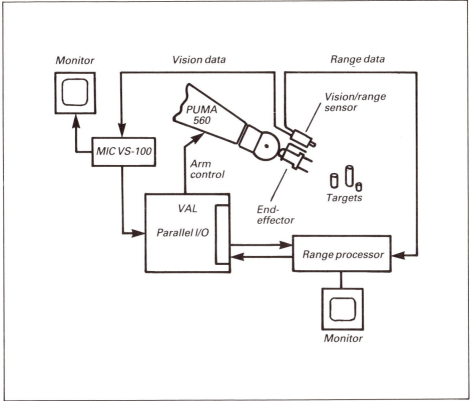

Fig. 6   Dynamic evaluation system

order to address the issues just discussed, some targets were fabricated and software was written to:

- Identify and locate targets placed randomly in the robot work volume using vision data.
- Position the TCL on the targets based on the x, y location calculated by the vision algorithms.
- Range to a target both discretely, at a fixed robot arm position, and continuously as the arm moved towards a target.
- Use the preceding techniques for target identification and acquisition.

Initially, one-inch-diameter cylinders varying in height between one and five inches were fabricated for targets. The system was then exercised in the following fashion to determine where specific problems existed. The cylinders were placed randomly on the table beneath the robot. The robot arm moved the sensor to a viewing point above the table and a picture was taken. The x, y location of each cylinder was determined based on the vision data. The sensor was then positioned approximately 30cm above the table and a discrete range point was taken to facilitate differentiating height. The sensor was positioned such that the TCL field of view was incident on the front edge of the cylinder. The cylinders were then acquired by approaching along the acquisition axis while monitoring range, and stopping the motion when the range reached an appropriate threshold. This procedure was performed several times and the following observations were made:

- The positioning of the TCL field of view on the target varied significantly from cylinder to cylinder.
- The sensor had trouble differentiating between cylinders that varied in height by less than 1.5cm.
- The placement of the gripper fingers on the cylinders upon acquisition varied from cylinder to cylinder.

In general, this performance was not as expected based on the static performance analysis. These effects were consequently analysed in greater detail by observing the individual functions of the process, i.e. vision x, y location, discrete ranging and range monitoring. The following conclusions were drawn.

X, Y positioning. The variance in positioning of the TCL field of view on the targets was caused by vision parallax. This can be explained as follows. The vision processor was trained with three-inch cylinders, and the transformation relating camera location to robot location was based on this training. When other than three-inch cylinders were viewed off-axis (not directly below the sensor) the calculated x, y position was closer or further from on-axis for the one-inch or five-inch cylinders respectively. This problem was helped somewhat by viewing higher above the table, but positioning variations were still evident. This problem was aggravated by the large height difference between the cylinders.

Discrete ranging. In some cases it was evident that the TCL field of view was not only positioned on a cylinder edge, but also on additional features in the background. In most cases these additional or 'ambiguous' features were shadows. Recall that the enhanced algorithm uses the entire signatures in the pixel shift calculation. Therefore, the calculated pixel shift would be an average for the range to the shadow and the range to the edge. Observation of this phenomena brought to bear the necessity to be able to choose the portion of the signatures which were used in the pixel shift calculation, facilitating ranging to features of interest. This would be especially true, for example, when ranging to small parts grouped closely together. Therefore, some additional experiments were performed to determine if the TCL signatures could be segmented prior to calculating the pixel shift.

It was noted by examining the signatures that the intensity changes in the shadow region were very gradual as compared with those in the edge region. The logical choice then for segmenting was a slope thresholding technique, i.e. only the portion of the signature with large intensity changes would be used. A slope thresholding technique was implemented for analysis. The technique worked well; it could reliably segment the edge information in the TCL signatures, and facilitated a pixel shift calculation that was not influenced by background features.

Acquisition variance. After observing the acquisition process extensively, it was determined that there were significant variations in the shape of the TCL signatures depending on the TCL positioning as that target was approached. The positioning variations were due to the parallax problem discussed earlier. The difference in signature shape can be explained as follows. When the TCL is interrogated for data, a comparator in the TCL interface circuit compares the average light level incident on the TCL detectors with a setting called the AGC. The TCL detectors are allowed to collect light until the average reaches the AGC setting. The intent of this feature is to allow for operation in varying light levels. However, in the case of approaching the cylinders, an adverse effect occurs. Because at close ranges the cylinder is approximately the same size as the TCL field of view, the average light level varies significantly with TCL positioning. In some cases the TCL field of view is primarily on the background, and the light intensity on the cylinder is high with respect to the average light level. In this case the A and B signatures show a large intensity change across the cylinder edge. In other cases the TCL field of view is primarily on the target, and the light intensity on the cylinder is not much higher than the light level. In this case the intensity change across the cylinder edge is reduced. This has the effect of causing signature slope variations with varying TCL position. As the enhanced pixel shift algorithm is a function of slope, this can change the calculated pixel shift, and therefore the calculated range, as the TCL positioning varies.

To address this problem some data were taken where the AGC was adjusted manually at different positions to 'normalise' the signatures. In general, by adjusting the AGC, the signatures could be made to look similar at varying positions. In addition, this seemed to cause the calculated pixel shift to be more invariant. These results indicate that the positioning problem could be tolerated if a normalised set of signatures could be maintained. A solution would be to use an iterative approach to obtain the TCL data. To do this the AGC could be defeated, and the integration time could be software controlled to stop integration when the data met some specified criteria.

To further exercise the sensor dynamically and to get some data using small parts, an additional analysis was performed using small washers. The washers were approximately 1cm in diameter with 0.16cm walls, and had varying thicknesses. A similar analysis was performed using the same system where two washers were placed on a surface randomly, identified and located using vision, distinguished based on range, and the larger part acquired using range information.

After conducting this analysis, it was found that the problems which caused limitations when working with tall (1–5in.) cylinders were not as evident with the small parts. Since the parts were very short, the shadowing could be controlled relatively easily. Also regardless of the TCL positioning on the target, there were never significant variations in the amount of the TCL field of view incident on the target and the amount of the TCL field of view incident on the background. This eliminated the AGC problem which caused acquisition positioning variations with the cylinders. At close ranges (10–15cm), washers varying in thickness by as little as 0.5mm were distinguished with a high degree of reliability. Although the exercise of working with tall cylinders brought to bear some potential problems requiring more robust algorithms and strategies, it was concluded that these problems would not be as significant when dealing with smaller parts.

Concluding remarks

The integrated vision/range sensor provides both 2-D vision and single-point ranging functions in a lightweight package. The sensor is applicable for unstructured robotic identification and acquisition tasks, and has potential for becoming a reliable and low-cost sensor. The range resolution at close ranges is competitive with, if not better than, existing single-point ranging techniques. This can facilitate distinguishing between parts with slight differences in geometry, which cannot be distinguished with vision alone.

Acknowledgements

Much of this research was performed under contract F33615-82-C-5098 with the Air Force Wright Aeronautical Laboratory, Wright-Patterson AFB, Ohio. This program was sponsored by the Defense Advanced Research Project Agency, Arlington, Virginia. The authors would like to thank the engineering staff at Honeywell's Visitronic Division for their contributions and suggestions during this programme.

PRECISE ROBOTIC ASSEMBLY USING VISION IN THE HAND

S. Baird
AT & T (Bell Laboratories), USA
and
M. Lurie
RCA Laboratories, USA

Adaptive strategies using robotic vision are often frustrated by arm inaccuracies. Precise mounting of imprecisely presented loudspeakers has been proven feasible using an array imager, compliance and force sensing placed in the gripper of a PUMA 600 manipulator. The precision required to place loudspeakers on the given mounting posts was challenging: ±0.5mm translation and ±0.5° rotation. Placing the camera in the gripper permitted a sequential servoed strategy in which image feature estimates were nulled to pre-taught ideal images for a pallet of representative poses. An extreme-wide-angle lens brought the camera close to the workpiece, further reducing the effect of arm inaccuracies. Standard binary image features, computed from global shape moments, were not sufficiently accurate: a simple structural analysis of local features was needed. Typically fewer than three vision-servoed iterations were required to effect a 100-fold reduction in location uncertainty. Vertical compliance, instrumented with strain gauges, permitted tactile sensing of correct and erroneous conditions during grasping and mounting motions. In the rare (5%) case that visual servoing failed, the small residual errors were adapted to using this tactile feedback.

A successful robotic assembly demonstration, requiring high precision, guided exclusively by vision and force sensors is described. The experience was interesting and in some ways surprising. Even with a precise six-axis robot (PUMA 600) and a modern vision system (from SRI), the combined constraints of arm reach, camera weight, lensing and system accuracy posed considerable problems.

Motivation

This work is part of ongoing research into small odd-form-factor parts assembly in support of advanced manufacturing groups throughout RCA. While we recognise the value of product design for automatic assembly, and the cost-effectiveness in certain cases of special-purpose fixturing, we believe there will always be a large residue of assembly tasks demanding adaptation to imprecisely located parts.

Our adaptive robotics project provides laboratory feasibility demonstrations of the advantages of software- and sensor-intensive automation: improved flexibility, generality and reliability, with minimal set-up and calibration effort. Other coordinated robotics efforts within RCA Laboratories focus on manufacturing systems analysis, design for assembly, and applications of off-the-shelf robots using conventional parts-feeding systems.

Choice of application

The selected task was to automate the assembly of electronic parts to a TV 'mask' (the plastic bezel surrounding the picture tube). There are over a dozen mask models manufactured, undergoing annual style changes. Under these circumstances, design for automatic assembly would demand strict coordination with a variety of vendors, and special hardware fixturing would be frequently scrapped. The most promising approach to automation of this labour-intensive area seems to be sensor-guided robotics.

We started with the task of mating the loudspeaker with mounting pins on the mask (see Fig. 1). We knew that the TV mask could be fixtured, but the loudspeakers would be imprecisely presented. Although the loudspeakers vary in size and shape among TV models, they are all circular or elliptical, with four small mounting holes in corner flanges, and they all possess a flat metallic surface above the centre of mass (see Fig.2). Thus, a single-gripper design, using an electromagnetic pick-up, could be used with the entire family of loudspeakers. Also, the presence of holes at the critical mounting positions invited the use of binary images, the fastest, most reliable array image acquisition technique in common use.

The loudspeakers come to the factory packed in loose cardboard pigeonholes. We planned to acquire and orient them in two stages. First,

Fig. 1   Mating a loudspeaker with a mounting pin – the second pin is hidden by the hand, upper right

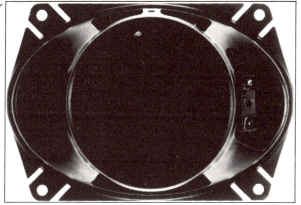

Fig. 2 Silhouetted loudspeaker showing mounting holes (two diagonally opposite holes are used)

they would be removed using a pick-and-place robot with a blind gripper and placed, roughly palletised, onto a light-table. Second, a sensor-guided assembly robot would locate them precisely, grasp them and place them on the mounting pins. Two mounting holes, diagonally opposed, would be slipped over two pins, one at a time. The second, sensor-guided stage is the subject of this report.

High-precision requirement

The holes in the loudspeakers are 5.5mm in diameter. The two mounting posts on the TV mask are 5mm in diameter, but rounded at the tip, so that the effective translational tolerance is ±2mm for successful mating. Since two holes must be fitted over two posts, and the posts are 148mm apart, there is an independent rotational tolerance of ±0.8°. Combining the two tolerances, we established the goal: locate the loudspeaker within ±0.5mm in the X- and Y- directions and ±0.5° rotation.

We estimated that a loudspeaker could be placed, using a simple means, onto a light-table with an error of not more than ±50mm of translation and ±50° rotation. Thus the sensor-guidance would have to reduce the uncertainty in position and orientation by two orders of magnitude.

We wished to be prepared in case vision failed for some combination of loudspeaker and mask, owing to manufacturing tolerances. The system was therefore designed to distinguish success from failure, and to be able to recover gracefully, to try again or to discard the loudspeaker.

Prior work

Albus et al.[1] have placed a camera in the gripper and demonstrated how hierarchical control fed by multiple sensors could support fast, complex adaptive behaviour. The accuracy of this system has not been reported to our knowledge.

Makhlin[2] has reported use of a hand-mounted camera to guide grasping of parts which were roughly palletised but otherwise randomly positioned and orientated. Only a single picture was used, and a '3-4s' assembly cycle was achieved by DMA block data transfers at 1Mbaud between manipulator and vision controllers. The accuracy is not described.

Asano, Maeda and Murai[3] have described the visual location of electronic component leads within ±0.12mm (worst case), down from an initial uncertainty of ±0.80mm (std dev.), in translation alone, to guide automatic insertion into PCBs. As in our application, insertion occurred one lead at a time. Two high-resolution (320 × 240) cameras in fixed positions were used.

Burgess, Hill and Pugh[4] have used a manipulator to position a camera so that the object image filled the field of view, for precise inspection purposes.

A clear and helpful discussion of some engineering trade-offs offered by placing a camera in the hand has been given by Loughlin and Hudson[5]. We share their interest in the large effective resolution obtained and the advantages of using extreme-wide-angle lensing to increase the depth of field.

Laboratory equipment

Although the scale of the task suited the equipment in our laboratory, the Unimation PUMA 600 and the SRI vision module, we found that the accuracy goals pushed both to their limits.

PUMA 600 manipulator

The Unimation PUMA 600 had a reach large enough to include a 2 × 3 pallet of loudspeakers as well as the TV mask. The repeatability of the arm, 0.1mm, was one-fifth of our accuracy goal. Its VAL controller permitted communication with a workcell controller (although somewhat awkwardly). Its six axes would permit the loudspeakers to be lowered onto the pins one at a time, and in case of failure would permit a dextrous hunt for the pin.

Unfortunately, the PUMA's accuracy is worse than its repeatability[6]. This is due to a mismatch between the software arm model and the actual arm, and is a common problem for articulated (as opposed to Cartesian) arm architectures. On our PUMA, we observed inaccuracies of up to 10mm over a 200mm straight-line path. We also observed rotational inaccuracies of as much as 0.9° in the last (sixth) joint of the PUMA. However, the arm was locally accurate: both errors were smaller for short incremental motions.

We did not expect to find a simple (e.g. linear) transformation from the visual field of the camera to the working space of the arm. One picture would perhaps not be enough, but a sequence of pictures and moves might converge rapidly.

SRI vision module

The SRI vision module, a laboratory prototype with the Agin binary image analysis software[7], has GE TN2200 solid-state 128 × 128 pixel cameras. Since the loudspeaker would be viewed lying isolated on a light-table, we were confident we could estimate its location to sub-pixel accuracy. This proved to be true for translation: when the loudspeaker occupied roughly two-thirds of the field of view, its location could be estimated to within 0.05mm.

However, we found that the estimated angle of loudspeaker orientation, computed using second moments of area within the SRI software, was not repeatable to less than a degree. This was due to the fact that the loudspeaker shape is not strongly elongated (see Fig. 2) (by contrast, for pencil-thin shapes we are able to measure angle to a tenth of a degree). We were forced to estimate the angle by means of a simple structural analysis: find all four holes within the silhouette; pair off the holes forming diagonals; decide which pair are the mounting holes; and compute the average angle using the centroids of the two pairs. This provided repeatable estimates to about a tenth of a degree.

Workcell control

The PUMA and SRI systems are controlled in our laboratory by a workcell computer (DEC PDP-11/34A running the multi-tasking operating system RSX-11M). Coordination programs are written in standard Pascal (compiler by Oregon Software), with syntax extended by us to express concisely the dialogue with the arm and eye.

Our policy has been to emphasise prototyping in Pascal on the workcell controller, to take advantage of its powerful program development environment. A disadvantage is that long dialogues with the devices go slowly (a command-reply pair can take 0.3s), owing to many layers of software and asynchronous serial transmission of ASCII strings. After feasibility has been demonstrated, these delays can usually be greatly reduced, by moving code from the workcell controller to the vision system, and by use of a more efficient computer-to-computer protocol with the arm (such as the 'ALTER/WALTER' extensions to VAL[6]).

Alternative approaches

We considered using a fixed camera above the light-table. The advantages were that the gripper design would be simpler, we could overlap vision processing and arm motion activity, and arm inaccuracies would not contribute additional uncertainties during calibration. However, the effective resolution of the system would be severely limited: experiments showed that to estimate the pose of the loudspeaker precisely, it must fill at least three-quarters of the field of view. Thus, no more than one pallet position could be covered by one camera. Also, we were not confident that we could model the arm inaccuracies well enough to permit precise guidance. If it turned out that we could not, the approach would have to be abandoned.

The second alternative, to mount the camera on the hand, relieved both disadvantages of the fixed-camera approach. High resolution, over a large field of view, would effectively be made available.

Also, even lacking an accurate model of the arm-eye system inaccuracy, we could expect that the good local accuracy of the system would permit a series of servoed moves, computed using a simple linear model, to converge. If convergence was fast, then we would have solved the problem without an arduous calibration of the arm. If it was prohibitively slow, then we could invest the effort to find a better model. If we found a highly accurate model, then only one servo step would do, and we would be not much worse off than with a fixed camera.

Of course, even at best, hand-mounted vision ties up the arm during image acquisition (with the present generation of cameras and arms). We judged that this lost time was out-weighed by the advantages and chose to mount the camera in the gripper.

Gripper design

A gripper for the PUMA robot was designed (Figs. 3 and 4), incorporating a camera, electromagnetic pick-up, vertical compliance, horizontal stiffness and vertical force sensing in a 1.4lb (0.63kg) package.

A GE TN2200 camera was disassembled and the camera/SRI electrical interface was redesigned to be lighter and smaller. About half the weight of the camera was in its strong cylindrical case. We replaced that with a less rugged but lighter aluminium box. The long cylindrical format also was awkward for mounting on the PUMA arm. That problem was solved, and the camera made more compact, by rotating the camera head (final circuit board, sensor array and lens mount) so that it was facing down, while the remainder of the boards were horizontal.

Two beryllium-copper plates, each $5.5 \times 2 \times 0.015$in., are used as springs, as shown in Fig. 4, to provide 10mm vertical compliance without rotational or horizontal compliance. The free end of the springs contains an electromagnet, controlled by a digital output from the VAL controller, to lift the loudspeaker. For vertical force sensing, four identical strain gauges are mounted symmetrically on the springs, two on top of the upper spring and

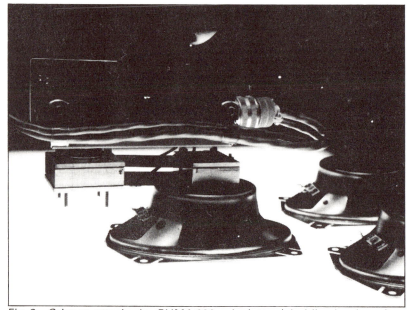

Fig. 3   Gripper, attached to PUMA 600 wrist (at top), holding loudspeaker

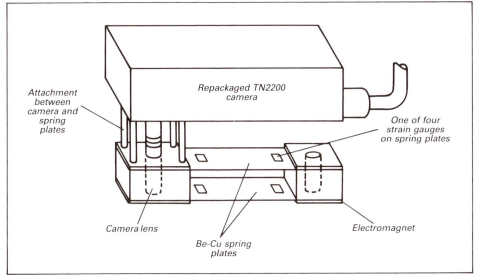

Fig. 4 PUMA gripper for picking up loudspeakers

two on the bottom of the lower spring. They are arranged in a bridge so that their outputs add. Since they are all thermally connected, and the bridge contains no fixed resistors, the system is very stable.

Operation of the system involves a one-time calibration, followed by repeated assembly steps.

Calibration

The tool offsets for both magnet and camera lens are directly measured and checked visually, by rotating about their centres, using VAL TOOL offset codes. Under program control, the fixtured loudspeaker mounting location on the pins is taught, and the pallet shape is roughed out by teaching a corner and two extreme points. Then, aided by a program, a loudspeaker is moved from the mounting pins to each pallet position in turn, and the operator touches up these poses so that the loudspeaker just touches the table and lies flat (this is another result of the non-linear distortions within the VAL world model).

Threshold choice. At this point, it is necessary to choose a threshold to produce usable binary images for pictures taken at a standard height above the light-table. This has been semi-automated: a rough manual choice is refined by a program performing a binary search among thresholds, guided by a simple model of the expected appearance of the loudspeaker. Careful adjustment of the camera and interface electronics and thorough shielding of the imager from stray light have permitted a usable threshold range of ±4%, which is enough to protect against drift in camera sensitivity and ambient light for several days.

Ideal images. Under the control of a third program, a loudspeaker is placed at each pallet pose and the gripper rises to the standard viewing height; the operator manually touches up the viewing pose to ensure that the loudspeaker is near the centre of the field of view and appears to the camera to rotate about its centre of mass when the arm is rotated about the lens offset. This touching up is designed to reduce dependencies on the gripper orientation (it could perhaps be automated). Then a picture is taken and the inferred position and orientation of the silhouette is reported. A file is written, recording, for each pallet position, the pick-up pose, the viewing pose and the image description. These are considered the 'ideal' prototypes for each pallet position.

Assembly

The assembly program reads the file of 'ideal' pallet data and proceeds to view each pallet position in turn. If no loudspeaker is seen it advances to the next pose. If part of a loudspeaker is seen (shifted partially out of the field of view) it will perform a large motion to bring it wholly in view.

Image nulling. The difference between the observed image and the 'ideal' image is used to compute a correction move, using a linear transformation that was estimated when the gripper was first installed. A series of pictures and moves is performed until the observed difference is less than the established precision goal of ±0.5mm translation and ±0.5° rotation.

Applying the offset. The offset vector from the ideal viewing pose to the final servoed viewing pose is computed using VAL operators. This is added to the ideal pick-up pose (again using VAL operators) and the loudspeaker is grasped at the adjusted pose.

Mounting on the pins. Finally, the loudspeaker is moved to the pre-taught mounting pose and lowered onto the pins, with the hand tilted so that one hole engages first. If the strain gauge sensors report a NEUTRAL height, the assembly to that pin has succeeded and the other hole is lowered. If the sensors report UP, then a tactile guided hunt is begun for the pin. The hole is lifted and the loudspeaker is pivotted about the other hole (on the assumption that it may be correctly positioned), alternately clockwise and counterclockwise, at greater and greater distances, until success or too great a displacement (the maximum is about 10 tries in each direction) is achieved.

Lens choice

A good deal of effort went into the choice of a lens. We first used a 16mm lens, at a viewing height of 390mm. In spite of careful manual calibration, we were frustrated by large translational errors for loudspeakers with large initial angular displacements.

It was hard for a while to see how this could occur. We performed more and more careful calibrations; we adjusted the camera electronics; we experimented with our own transformation mathematics, suspecting that the computational precision of the VAL operators was insufficient; we investigated barrel or pincushion distortion in the lens. None of these influences could be shown analytically to cause the symptom, and no attempted improvements along these lines relieved the problem.

Finally, we suspected uncontrolled orientation of the hand during calibration, especially rotation about coordinates in the horizontal plane

(VAL WORLD RX & RY angles). Variation of as little as 5° from vertical while acquiring the ideal images could, we argued, cause large translational errors later, when applying the visual offset. We could not be sure how severe this was, since we lacked a way to measure these angles independently of the VAL system.

After some deliberation, we chose to minimise the supposed effect by moving the camera closer to the table. Using an extreme-wide-angle lens of 5.5mm we could view the loudspeaker from a height of 134mm. The translation errors were greatly reduced, by more than the factor of three that might have been expected.

We were pleased with the performance of the extreme-wide-angle lens for several reasons. Shorter, quicker arm motions could be used. The depth-of-field, of course, was improved, simplifying the set-up. Parallax was not a problem in our case, since the mounting holes happened to lie in thin flat metal flanges some distance from the loudspeakers' central thickness, and so were not obscured. There was vignetting, but this did not trouble the connectivity software.

Results

We ran roughly 50 trials using the final version of the software. The visual guidance succeeded in 95% of the trials. Typically fewer than three servo steps were required. The tactile search successfully completed almost all of the trials in which visual guidance failed. In most cases only two tactile moves were required, but occasionally many were required.

Timing

Time for visual servo step averaged 12s. This was largely due to communications delays, described earlier. Simply by balancing computation among the existing equipment, servo time could be cut to about 3:1s on average for image acquisition and analysis, 0.5s for arm motion and 1.5s for communication among the devices.

If the visual servo time were optimised, the average assembly time would be about 12:6s for two servo steps, 4s to pick up and attempt to place a loudspeaker and 2s to mate it with the pins.

The time for a tactile servo step was about 1s. These occurred rarely enough that the average cycle time was increased by about 10%.

Concluding remarks

The virtual absence of hard fixturing in this application places in high relief some advantages and problems of sensor-based robotic assembly. Having the full six degrees of freedom in the manipulator both helped and hindered the effort: it permitted a dextrous tactile hunt but it complicated calibration and use of visual feedback.

It is evidence of the power of visual servoing that, in spite of low system accuracy and the impracticability of exact calibration, it was typically possible, in two servo steps, to reduce location uncertainty by a factor of 100, down to a small multiple of the underlying arm repeatability.

Acknowledgements

Merryll Herman wrote the bulk of the image processing and nulling code, enhanced the SRI vision module code and ran many of the trials. Sal Noto fabricated much of the gripper, including the electronics, helped repackage the camera and adjusted the camera and strain gauges. The paper has benefited from a critical reading by Eleanore Wells, Mike Carrell and John Aceti. The authors would like to thank Allen Korenjak and Art Kaiman for their support and interest.

References

[1] Shneier, M., Kent, E., Albus, J., Mansbach, P., Nashman, M., Palombo, L., Rutkowski, W. and Wheatley, T. 1983. Robot sensing for a hierarchical control system. In, *Proc. 13th Int. Symp. on Industrial Robots*, Chicago, pp. 14-50 – 14-66. SME, Dearborn, MI, USA.

[2] Makhlin, A. G. 1982. Vision controlled assembly by a multiple manipulator robot. In, *Proc. 2nd Int. Conf. on Robot Vision and Sensory Controls*, Stuttgart, pp. 83-92. IFS (Publications) Ltd, Bedford, UK.

[3] Asano, T., Maeda, S. and Murai, T. 1982. Vision system of an automatic inserter for printed circuit board assembly. In, *Proc. 2nd Int. Conf. on Robot Vision and Sensory Controls*, Stuttgart, pp. 63-67. IFS (Publications) Ltd, Bedford, UK.

[4] Burgess, D., Hill J. and Pugh, A. 1982. Vision processing for robot inspection and assembly. In, *Proc. Robotics & Industrial Inspection*, SPIE Vol. 360, San Diego, CA, USA, pp. 272-279.

[5] Loughlin, C. and Hudson, E. 1983. Eye-in-hand robot vision scores over fixed camera. *Sensor Review*, 3(1): 23-26.

[6] Carlisle, B., Roth, S., Gleason, J. and McGhie, D. 1981. The Puma/VS-100 robot vision system. In, *Proc. 1st Int. Conf. on Robot Vision and Sensory Controls*, Stratford-upon-Avon, UK, pp. 149-160. IFS (Publications) Ltd, Bedford, UK.

[7] Gleason, G. and Agin, G. 1979. A modular vision system for sensor controlled manipulation and inspection, A.I. Center Technical Note 178. SRI International, Menlo Park, CA, USA.

LINE, EDGE AND CONTOUR FOLLOWING WITH EYE-IN-HAND VISION SYSTEM

C. Loughlin
Electronic Automation Ltd, UK
and
J. Morris
Unimation (Europe) Ltd, UK

This paper demonstrates by application some of the possibilities and major benefits of eye-in-hand robot vision for line, edge and contour following, height sensing and dimension checking. In many cases operations that are not possible or not cost-effective using other approaches become almost trivial in their simplicity using these methods. The I-SIGHT 32 vision system and Unimation robot combine to tackle tasks in a way that makes the programming of a particular application very easy and also highly tolerant and adaptive to the mechanical deviations that are always present in the manufacturing environment.

The very great benefits of using eye-in-hand robot vision as opposed to static overhead camera systems have alrady been discussed elsewhere[1, 2] in relation to hole location, pick and place and jig location tasks. This paper considers applications in line, edge and contour following, height sensing and dimension checking.

The applications use the I-SIGHT 32 vision system with spot and 'line of light' lasers and a Unimation PUMA 560 robot running under VAL II and using the Alter/Walter high-speed external path control function (Fig. 1).

System overview

The basic operation of the system is that the camera takes in visual information, processes this picture to determine either position, distance or orientation information, and then instructs the robot to move such that, for example, either the position is centralised, a constant height is maintained, or orientation is corrected.

The most important feature of the system is that steps are taken to maintain a stable reference situation. For example, if a hole is located in the top right-hand corner of the camera's field of view, then the vision system will instruct the robot to move right a bit and up a bit in iterative procedure in such a way as to centre the hole within the field of view. Once the hole is centred, the robot then knows where it is. Until it is in the centre the system does not consider any further actions other than the operation of centering.

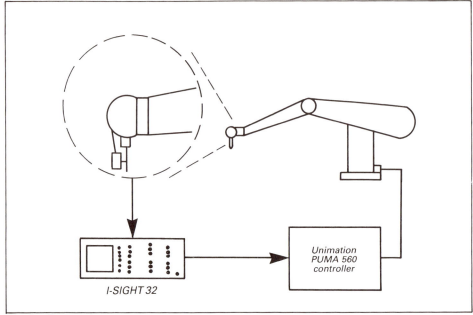

Fig. 1   System layout

Thus the system never considers the action "I want to insert a rivet at x location 15, y location 8", but instead first centres and then undertakes the action "I want to insert a rivet in the hole above which I am now centred". The distinction between these two processes is not obvious but it is important as a means of obtaining high accuracy and good performance.

The movement commands are sent to the robot via a high speed (19.2kBaud) RS423 serial link and the commands are transmitted every 28ms (36Hz). This high speed enables smooth and rapid motions to be achieved and even permits the tracking of a moving object. The communication protocol is specified by Unimation as part of the VAL II real-time path control function. This has been covered in an earlier publication[2], and also more completely in the Unimation 'Users Guide to VAL II'.

Real-time path control

There are two basic modes of operation for the real-time path control function (ALTER) of the Unimation PUMA. These are cumulative and non-cumulative. In cumulative mode, the effect of any data received is accumulated and the robot location is modified by the sum of all past alter data. Consider, for example, that the robot program is directing the robot to stay at a location and the external computer sends an X alter value of 0.1mm in response to each alter data request. The result will be that the robot will move away from its nominal location in the X direction at a rate of 0.1mm per 28ms (a speed of 3.5mm/s). Furthermore, if the external computer then sends an X alter value of zero, the robot will stay at its last location. (The applications covered in this paper all use the cumulative mode.)

In non-cumulative mode, the robot location is modified only by the most recent data. Then, if the external computer sends an X alter value of 0.1mm in response to each alter data request, the robot will move away from its nominal location 0.1mm in the X direction and stay at that offset location. If the external computer then sends an X alter value of zero, the robot will return to its nominal location.

Visual feedback can be used in two modes which might appear to be different in operation but which are, in fact, identical. Two examples are used to illustrate these modes. In the first, which might be edge or contour following, the robot has basically been told to move from A to B. It uses visual feedback to maintain a constant distance and compensate for variations in the expected path. As the robot is moving from A to B cumulative alter data (Dx, Dy) that it received will cause an immediate displacement of the robot arm by Dx, Dy and also modify the position of the target location B by the same amount.

In the second example, which is used in applications such as hole location, the points A and B are at the same place, i.e. the robot is stationary. Cumulative alter data that is received will therefore cause an immediate displacement of both the robot's position and the location of point B (which are the same thing) by the amount Dx, Dy. Thus the processes of path and position modification are identical in operation.

The camera system

Two types of optical sensors are to be described. Both use lasers and both only detect the physical, and not optical, characteristics of the part in view. Indeed, the camera does not see the object at all – simply the laser light reflected off the object's surface. This has a number of benefits: the first is

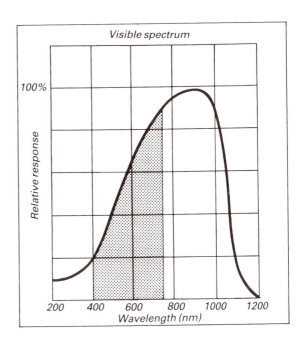

Fig. 2   Spectral response

that the colour, light intensity or visual contrast of the picture becomes completely irrelevant and the same information can be derived off black, white or grey surfaces. The second advantage is that we are measuring the physical location of the edge/surface and not the rather incidental occurrence of a change in light intensity. As it is almost always the physical boundaries that we wish to detect, it is obviously better that they be obtained 'first hand'.

Single-spot laser probe

The use of the miniature camera in conjunction with a collimated solid-state laser enables the vision system to act as a ranging device. The laser is a solid-state continuous wave 2mW laser (Mullard CQL13A) operating in the near infrared at a wavelength of 820nm which coincides approximately with the peak sensitivity of the camera (Fig. 2). An infrared filter (Kodak 88A) is used within the camera in order to attenuate the effects of illumination in the visible spectrum.

The laser is mounted vertically with the camera adjacent to it but at an angle to the laser axis which is set according to the range and depth of field that is required (Fig. 3).

The laser projects a dot of light onto the surface and it is this dot that is picked up by the camera. The picture seen by the camera will contain the single blob, and it is the vertical position of the blob that defines the distance of the laser from the surface. If the blob is in the vertical centre of the picture then the surface will be at a distance of r. It should be noted that the vertical position of the blob is not linearly proportional to distance but this does not materially affect the operation of the system as the feedback loop will always act to keep the robot at distance r from the surface.

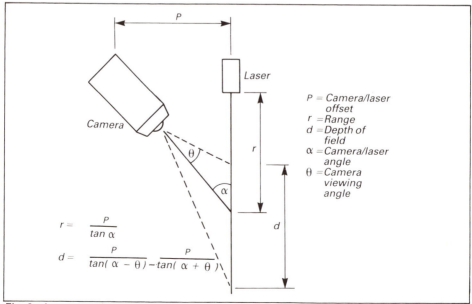

$$P = \text{Camera/laser offset}$$
$$r = \text{Range}$$
$$d = \text{Depth of field}$$
$$\alpha = \text{Camera/laser angle}$$
$$\theta = \text{Camera viewing angle}$$

$$r = \frac{P}{\tan \alpha}$$

$$d = \frac{P}{\tan(\alpha - \theta)} - \frac{P}{\tan(\alpha + \theta)}$$

Fig. 3   Laser ranging system

The parallel beam projected by the laser is 5.4mm in diameter which is large by collimated laser standards. This is important, however, in view of the increased accuracy that is made possible in the determination of the centre of the beam (Fig. 3).

The low resolution of the camera also requires that a broader beam be used as it would otherwise be possible for the laser light to fall not on a photosite but rather on the warp and weft of the silicon structure and thereby not show up on the camera image.

Applications of the single-spot laser probe

Contour following/height sensing

If we take an example where the robot is required to lay a track of glue in a straight line across a curved surface, without height sensing it would be necessary to teach the robot a large number of points across the surface so that a constant height could be maintained. Typically, this might take in excess of an hour to program, would be prone to operator error and would have to be repeated for all other glue paths across the surface.

Using the height-sensing feedback from the vision system, it is only necessary to teach the start and finish points to the robot as all height variations would automatically be detected by the height sensor and the robot path would be modified accordingly. This would cut programming time down to about 5min, eliminate operator error and also compensate for variations in the position of the surface caused by jig tolerances.

Object location/dimension checking

In this application example, we wish to check certain dimensions of a welded frame structure. The vision system is low resolution and therefore cannot be used by itself to obtain precise measurements. However, a robot such as the Unimation PUMA 560 has a resolution of 0.1mm and can therefore be used effectively as a measuring instrument can sense what it is measuring.

Assuming that the welded frame is a simple rectangular construction and that we wish to take a number of key measurements (Fig. 4) in order to determine its quality. The process of obtaining each measurement is that the robot starts with the single-spot scanner pointing at free air and then moves in towards the expected location of the edge and keeps moving until the vision system detects the edge. The robot will then servo under visual feedback until the laser dot is in the centre of the camera's picture. Once this has been achieved, the robot then knows the precise location in three dimensions of the edge of the frame. Further measurements can be undertaken in the same way until all relative dimensions of the frame are known and can therefore be used to determine whether the position, size and squareness of the frame is to within manufacturing tolerances. These calculations can all be undertaken by the robot control system and an accept/reject decision made.

The key to this application is that the measurement system is the robot itself with the vision system acting solely as a sensor or guide that tells the robot when to take a measurement.

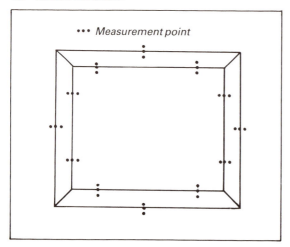

Fig. 4   Dimension checking

Line of light laser probe

Much work[4] has already been done on the detection of edges using a projected line of light and a camera system that analyses the angle and curvature of the projected line and detects discontinuities in it which represent the edge or border in question (Fig. 5). This work has primarily concentrated on the automatic following of edges and seams during welding operations. This paper will aim to identify new application areas for the line of light laser techniques.

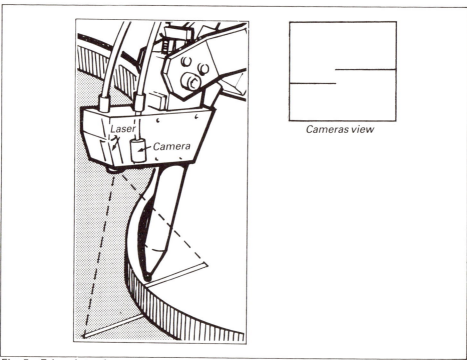

Fig. 5   Edge detection

The line of light is generated by a laser and can be produced either by the use of a cylindrical lens (e.g. a glass rod) in front of a collimated laser beam or more conveniently by using the direct output of a rather special solid-state continuous-wave laser whose output beam diverges at 50° in one axis but only 6° in the other so that it effectively projects a diverging line of laser light onto a surface. The particular device used in our research (STC LCO6-03) is a 6mW laser with peak emission in the near infrared at 850 nm. One major advantage of this line of light laser probe is its small size (about the same as a cigarette lighter) and its low weight which means it can be used in very confined or restricted areas.

The single-spot laser probe enabled surface measurement in one axis to be made (i.e. distance). With the line of light laser it is possible to determine surface angle from the tilt of the perceived line and also detect the position of edges more rapidly than is possible with single-point inspection.

Edge following

If the line of light is projected so that it runs orthogonally to the edge (Fig. 5) then the camera will only see the part of the line that hits the top surface with the end of the line marking the edge. The vision system processes this picture and then sends offset information (normally in just one axis) to the robot in order to keep the end of the line (the edge) in the centre of the picture. If the surface also undulates up and down this will show up as a vertical displacement of the line. This height information can be fed back to the robot in a similar manner to the contour following example.

In this way it is therefore possible to accurately follow an edge in three dimensions using very simple, cost-effective hardware.

Object location

In another example let us assume that cylindrical objects are presented to the robot in pallets but there is a variation in the absolute location of the cylinders caused by jigging and packing tolerances. The cylinders are very delicate and must be accurately located (±0.1mm) before they can be picked up by the robot's gripper.

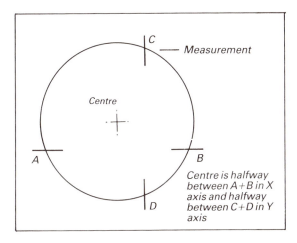

Fig. 6   Cylinder location

A high-resolution overhead vision system might well be able to identify the position of the cylinders but would not be able to offer sufficient accuracy of location due to parallax errors, optical distortion, scaling errors and mismatch between the camera's and the robot's coordinate systems.

By using a line of light laser in a similar manner to the previous edge following application it is possible to locate precisely four positions (two in X axis and two in Y) that will enable the centre of the cylinder to be determined accurately (Fig. 6). The principle limiting factor on accuracy in this application will be set by the robot rather than the vision system, for the same reasons as discussed in dimension checking.

Concluding remarks

This paper has demonstrated how the correct application of very simple techniques using an eye-in-hand robot vision system can provide a high-speed, cost-effective solution to many industrial problems.

The use of lasers eliminates the usual problems of grey-scale image processing systems and also means that the physical and not optical properties of an object are measured.

The principle of centring and maintaining a particular condition (i.e. laser dot in centre of picture) contributes greatly to the accuracy that can be achieved and, indeed, only limits this accuracy to that of the robot iself – which in the case of the PUMA 560 is very high.

References

[1] Loughlin, C. and Hudson, E. 1982. Eye-in-hand robot vision. In, *Proc. 2nd Int. Conf. on Robot Vision asnd Sensory Controls*, pp. 264-270. IFS (Publications) Ltd, Bedford, UK

[2] Loughlin, C. and Morris, J. 1984. Applications of eye-in-hand vision. In, *Proc. 7th Annual British Robot Association Conf*, pp. 155-164. BRA, Bedford, UK

[3] Hill, J. et al. 1978. Machine intelligence research applied to industrial automation. Eighth report NSF grant April 1975 – 13074. SRI project 4391, SRI International, Menlow Park, California.

[4] Clocksin, W. F. and Davey, P. G. 1982. Progress in visual feedback for robot arc-welding of thin sheet steel. In, *Proc. 2nd Int. Conf. on Robot Vision and Sensory Controls*, pp. 189-200. IFS (Publications) Ltd, Bedford, UK.

A NOVEL SOLID-STATE COLOUR SENSOR SUITABLE FOR ROBOTIC APPLICATIONS

P. P. L. Regtien and R. F. Wolffenbuttel
Delft University of Technology, The Netherlands

A novel solid-state colour photodetector is introduced with the capability to determine both the average colour and the intensity of the incident light within the visible part of the spectrum and without additional colour filters. This colour determination feature provides an industrial robot with an extra capability for environmental data extraction. Unlike an already developed silicon colour sensor, which is based on a vertical dual-diode structure, this new type of colour sensor consists of two adjacent photodiodes with different sensitive areas. In the practical colour sensor the reverse voltage ratio across the photodiodes can be adjusted until the imbalance in photocurrent, caused by the photodiode sensitive area inequality, is compensated by the width of the corresponding space charge region. At equal photocurrents the reverse voltage across one of the diodes is a direct function of the area ratio between the photodiodes, a reverse reference voltage across the other diode and the colour of the incident light. The colour response is not affected by either the wavelength dependence of the reflection coefficient or the quantum efficiency.

Several trends that can currently be perceived in industrial automation will lead to an increasing application of sensors in industrial robots in general, and in dexterous experimental assembly robots in particular[1]. From the sensor point of view these assembly robots are the most sophisticated types because of the multitudinous implementation of sensors for motion control and environmental explorative sensors that are needed for feature extraction.

The implementation of the first category of sensors is a direct result of the general tendency, occurring for reasons of economy as well as advantages in mobility, to turn the trade-off between the mechanical accuracy of a joint in a robot construction and the application of position sensors in favour of the latter. This development, in which the customary and expensive precision fittings are superseded by appropriate sensors, will result in a relatively supple robot construction, which features a large mobility while still maintaining a high position accuracy.

Furthermore a large interest is notable in flexible assembly robots suitable for relatively smaller series of automated production, which are preferably also equipped with an increasing decisive ability based on sensor information. This trend, occurring by virtue of both safety[2] and economy[3], resulted in the development and implementation of an entirely different class of sensors necessary for feature extraction. In current research, there is a widespread interest in sensors of this second category, such as vision systems and tactile imaging sensors. To achieve the above-mentioned increasing flexible insertability of the industrial robot, it could be advantageous to increase the diversity of these environmental explorative sensors.

Another relatively unknown type of sensor in this category, which is able to contribute toward providing information sufficient to achieve a successful mounting, consists of sensors based on colour perception and thus able to distinguish between identical objects of different colour. This can be performed by a two-dimensional colour imager, but it could be advantageous to avoid image processing by implementing this colour perception on a lower sensor level.

Such a solid-state colour sensor can perform that task because it is able to extract colour information of the incident light and could therefore contribute to a better feature extraction in some applications.

Semiconductor colour sensor

To obtain a correctly functioning colour sensor, it is necessary to compensate for other optical parameters such as brightness and quantum efficiency. In this sensor the compensation is performed in a natural way.

The operation is based on the exponential relationship between the absorption coefficient and the wavelength of incident photons with an energy inbetween 1.1 and 3.5eV, due to the indirect bandgap of silicon, so that the colour filtering is performed by the semiconductor itself. An already

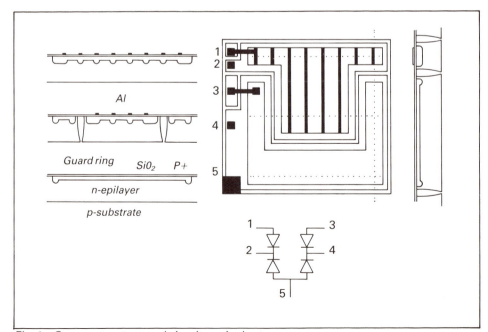

Fig. 1   Sensor structure and circuit equivalent

developed silicon colour sensor utilising this principle[4] is based on a vertical dual-diode structure, in which the photocurrent ratio between a shallow and a deep photodiode represents the colour. The new type of colour sensor, however, basically consists of two adjacent photodiodes with different sensitive areas. The basic sensor structure is shown in Fig. 1. The dimensions are $2000 \times 2000\mu m$. The two photodiodes extend over equal areas to reduce the effect of the dark current on the colour response. The sensitive area inequality is obtained by an unequal shielding of the photodiodes with aluminium surface strips or a difference in attenuation of the photocurrents in the signal conditioning circuits. The two diodes are interlaced, because a small light spot, which doesn't cover both diodes entirely, would otherwise cause a change in the effective sensitive area ratio and would therefore affect the colour response.

The photodiodes are formed by a boron implantation layer and the n-epilayer with the junction at a $0.5\mu m$ depth. The p-implantation top-layer is surrounded by a p-diffusion guard ring to increase the breakdown voltage. A reverse voltage applied across this junction will result in a depletion layer having a width equal to:

$$d = \sqrt{\frac{2\varepsilon}{eN} \; (V_d + V_{rev})}$$

where ε is the dielectric permittivity, $\varepsilon r(Si) = 12$, N is the doping of the lighter doped layer $[m^{-3}]$, and V_d is the diffusion voltage [v].

Due to the high doping of the top-layer, the space charge region will almost entirely be extended into the n-epilayer. The imbalance in photocurrent, caused by the photodiode sensitive area inequality, can be compensated by a decrease in the larger photodiode reverse voltage, which narrows the space charge region of the diode and thus approaches a current balance. The reverse voltage across the larger photodiode can be adjusted using a feedback loop until the photocurrents are in balance. After obtaining stationary conditions, the reverse voltage across the larger photodiode, which is a direct function of the smaller photodiode reverse reference voltage, the area ratio between the photodiodes and the colour of the incident light can be monitored continuously. The value of the photocurrent depends on the intensity of the incident light and the quantum efficiency of the photodetector.

It is necessary to prevent charge carriers generated beyond the depleted region, but within the diffusion length, from contributing to the total photocurrent. These charge carriers can be removed by reverse biasing of the substrate–epilayer junction, which should deplete the remainder of the epilayer. The epilayer–substrate reverse voltage V2 is directly controlled by the bias voltage of the reference diode, which is manually set to the appropriate value, and by the width of the epilayer. The bias voltage V4 depends in a similar way on the voltage measured in the colour loop mentioned before, according to the relation:

$$\sqrt{V_{d1}+V_{ref}} + \sqrt{\frac{N_s}{N_c+N_s}}\ (V_d + V_2) = W_c \sqrt{\frac{e\ N_c}{2\varepsilon}}$$

where N_c is the doping of the epilayer, N_s is the doping of the substrate, and W_c is the width of the epilayer.

The actual width of the epilayer can be determined by using a preliminary measurement in which the reverse voltage across a junction is increased until the entire epilayer is depleted. Such a measurement can be performed on both the epilayer–substrate junction and the top-layer–epilayer junction, through which the depletion voltages $V_{depl\ 1}$ and $V_{depl\ 2}$ respectively are obtained. The value of the doping ratio might reveal a large difference when compared to the nominal value, but it can be shown that this ratio equals the depleting voltage ratio:

$$\frac{N_s}{N_c + N_s} = \frac{V_{d1} + V_{depl1}}{V_{d2} + V_{depl2}}$$

With this result, the ultimate equation, from which the voltage required for depleting the remainder of the epilayer can be derived, now reads:

$$V_{d1} + V_{ref} + \sqrt{(V_{d2} + V_2)\left[\frac{V_{d1} + V_{depl1}}{V_{d2} + V_{depl2}}\right]} = V_{d1} + V_{depl1}$$

This equation can be solved continuously using a microprocessor system in the loop as shown in Fig. 2.

Unlike the human eye, this sensor determines the average colour of the incident light within the visual to near infrared spectrum ranging from 400 to 1000nm. The observed response is therefore not identical to that of the human eye. This could imply an essential limitation for robotic applications, because the objective is to obtain a uniquely determined colour representation of an object suitable for machine analysis. In principle it is possible for an illumination with two different colours to result in the same sensor response when only the average colour is detected.

This ambiguity can be avoided when the spectral standard deviation is measured as well. In most robotic applications, however, the colours to be detected are clearly distinguished, so the satisfaction of this unique determination condition is not expected to be too difficult to fulfil.

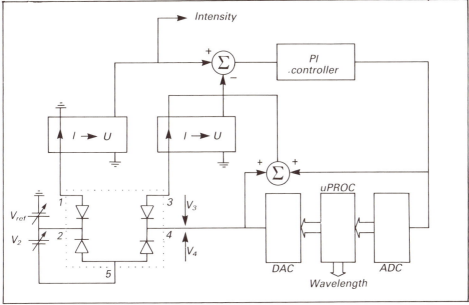

Fig. 2   Simplified biasing diagram of the sensor

It should be emphasised that, due to the current equality condition, the colour response is not affected by either the wavelength dependence of the reflection coefficient or the quantum efficiency of the photodetector. Furthermore, this sensor exhibits the particularly convenient characteristic of having a geometric degree of freedom left for the design of the colour range. To a certain extent this sensor reveals a 'low-pass' behaviour that can be affected by a proper choice of the area ratio and the reference voltage.

Experimental results

The colour response of the above described sensor can be simulated using a simple model, which calculates for each wavelength the corresponding absorption coefficient in silicon and the width of the depleted region necessary to obtain a current balance in a specified sensor configuration. The results of the simulations using an area ratio equal to 1.2 and 1.6 are depicted in Fig. 3 and Fig. 4 respectively, and clearly show the effect of the area ratio. Experiments have been performed using an already realised sensor and the results are depicted in corresponding figures indicated by the squares. The figures show that these experimental results are in reasonable agreement with the theory down to a wavelength of about 600nm at an area ratio equal to 1.6 and down to 500nm when using an area ratio of 1.2. The calculated reverse voltage that should be applied at a lower wavelength is not practical. The lower limitation of the range to 450nm at an area ratio close to unity is mainly the result of the large absorption of blue light within the p-implantation top-layer and the relatively small lifetime of the minority charge carriers in this heavily doped region. In further research, experiments will be performed using a more convenient doping profile to enlarge the spectral range to below 400nm.

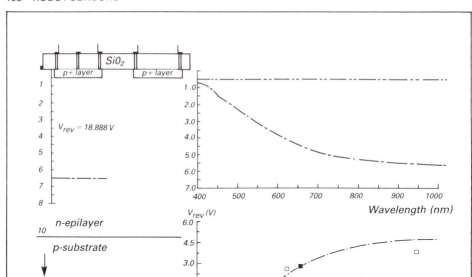

Fig. 3   Simulated and measured colour response at an area ratio of 1.2

The colour response should not depend on the intensity of the incident radiation. Such a demand can only be satisfied as long as the photocurrent is sufficiently large to determine the current balance. An experiment is performed in which a sensor is illuminated with an increasing intensity by a

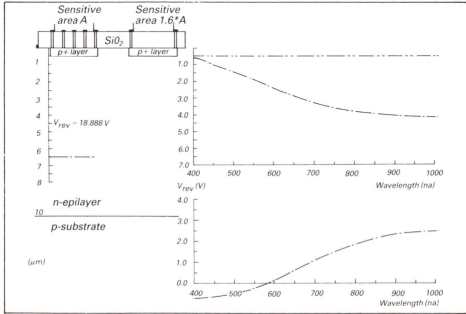

Fig. 4   Simulated and measured sensor response at an area ratio of 1.6

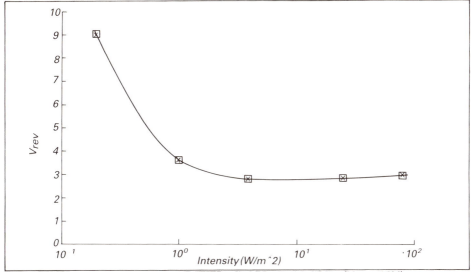

Fig. 5   *Colour response to a variable intensity of a fixed colour (V_{ref} = 10V)*

LED light source with a peak wavelength at 670nm. At a very low intensity, the current balance will solely be established according to the dark current ratio. Since the photodiodes are of the same dimensions, the reverse voltage, required to obtain the current balance, necessarily equals the reference voltage. This reverse voltage is, naturally, independent of the source wavelength. The results are depicted in Fig. 5, which shows the uncertainty at an intensity of less than 1 W/m². This result is in accordance with the obvious fact that a colour can only be perceived if an illuminating source is present.

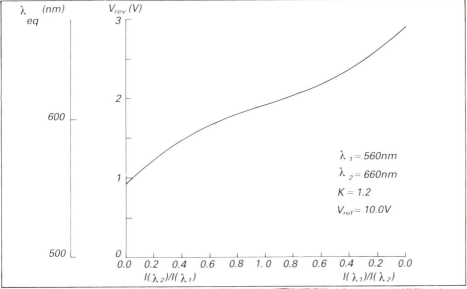

Fig. 6   *Simulation of the response to mixed colours with a variable relative intensity*

The simulated sensor response to illumination by two light sources of different colour and intensity is shown in Fig. 6. This figure illustrates that the sensor determines the average colour of the incident light. The sensor reveals a slightly larger sensitivity to the smaller wavelength due to the absorption of light of the large wavelength beyond the epilayer.

Application of the colour sensor to robotics

A typical problem in robotics, where the sensor described here can be of value, arises at the manipulation of objects that are identical except for a clearly distinct colour. In such a situation, vision systems and tactile imaging sensors will not provide sufficient information to select the correct object, whereas a colour vision system might be somewhat abundant, not to mention the inconvenience if the industrial robot is already equipped with an ordinary vision system. An experiment is performed, which demonstrates the suitability of this colour sensor in such circumstances.

In the experiment the sensor is moved along a number of wooden cubes, which are identical except for the colour, and meanwhile the sensor response is plotted. The colours of the cubes are yellow, green and red, respectively. The corresponding sensor response is depicted in Fig. 7, which indicates the detected colour when moving along the scene.

The scene is front-illuminated and the background is poorly reflective; therefore the colour response when positioned in between two cubes is determined by the dark current. As mentioned before, the reverse voltage will in that case be equal to the reference voltage. The measured colour reasonably fits the actual colour after conversion using the gauge-curve drawn in Fig. 3. An industrial robot can therefore distinguish between the cubes using this simple, solid-state colour sensor.

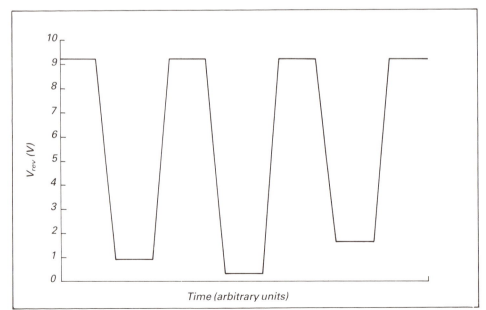

Fig. 7   Sensor response when moving along the scene

References

[1] Pugh, A. 1982. Second generation robotics. In, *Proc. 12th Int. Symp. on Industrial Robots*, pp.18. IFS (Publications) Ltd, Bedford, UK.

[2] Hasegawa, Y. and Sugimoto, N. 1982. Industrial safety and robots. In, *Proc. 12th Int. Symp. on Industrial Robots*, pp. 9-15. IFS (Publications) Ltd, Bedford, UK.

[3] Warnecke, H. J., Schweizer, M. and Abele, E. 1979. Cleaning of castings using sensor-controlled industrial robots. In, *Proc. 9th Int. Symp. on Industrial Robots*, pp. 535-544. SME, Dearborn, MI, USA.

[4] Kako, N., Tanaka, N. and Suzuki, C. 1983. Combustion detection with a semiconductor colour sensor. *Sensors and Actuators*, 4: 655-660.

3

Fibre-Optic Sensors

Optical fibres can be used in interesting ways in robotics. Images can be 'coupled' to a remote camera from robot grippers and fibres can be used singly for specialist purposes. Therefore, in this section the concept of 'distributed vision' needs to be borne in mind.

ROBOT EYE-IN-HAND USING FIBRE OPTICS

A. Agrawal
GCA/Industrial System Group, USA
and
M. Epstein
Northwestern University, USA

For many applications, eye-in-hand robot vision has several advantages over the use of a fixed external camera. Even smaller solid-state cameras are too bulky for gripper mounting and, therefore, limit the gripper's movement for many operations. A novel approach for eye-in-hand robot vision is proposed utilising coherent fibre-optic bundles for carrying light from an object to be imaged, and photodiodes to convert this light into electrical signals for processing. The feasibility of this concept has been experimentally demonstrated by developing a fibre-optic eye-in-hand system, mounting the scanner head on a gripper and obtaining good images under back and front light conditions.

The next generation of industrial robots will require adaptation of motion control based on feedback from the environment via visual, tactile and other types of sensors. This will allow these future robots to accommodate changes in workpiece position/orientation and perform complex operations such as assembly. The required sensors can be grouped in three areas: those needed prior to contact (vision, range, proximity), during contact (touch, slip) and after contact (force, torque). Amongst these, vision systems for robots have attracted the greatest research effort. A number of vision systems are now commercially available. The majority of these systems use a fixed overhead camera located above the work area.

While these systems serve a useful purpose by providing a global view of parts in the work area, they have several disadvantages[1]:

- The field of view of the vision system is obscured by the robot arm prior to the part being retrieved.
- Higher camera resolution is required for the same object resolution because of the camera distance from the object which also contributes to parallax error.
- Time-consuming calibration is required to establish robot to camera coordinate reference.

All of these problems can of course be solved simply by mounting the camera on the gripper. Many experimental systems have been built and have shown impressive results[2-4].

However, these systems are not yet able to provide visual information adequately for complex operations, e.g. precision assembly of small parts. Some of the shortcomings are:

- The cameras are not small enough to allow physical integration with the gripper and are simply mounted on the side with some fixed offset.
- Even with small CCD cameras, the weight on the arm due to the camera and the cable is enough to degrade the robot's performance.
- The processing of the image in real-time requires significant computing power.

In this paper an approach that can overcome most of these problems is described.

Basic concept

Our concept of a vision sensor is based on combining two technologies: coherent fibre-optic bundles for carrying light from an object to be imaged, and photodiodes to convert this light into electrical signals for processing. In a coherent fibre/optic bundle, each fibre maintains the same physical orientation with respect to all other fibres throughout the length of the bundle. This approach has been applied in developing flexible endoscopes for medical instrumentation[5]. In our concept, the light from the picture elements (pixels) carried by the fibres is converted into electrical, computer-compatible signals using photodiodes, which are available as single elements, as linear arrays or as matrix arrays. This concept is illustrated in Fig. 1, which shows a fibre-optic scanner head over a workpiece. The workpiece may be optionally illuminated by light delivered to the area of interest by another fibre-optic cable from a light source.

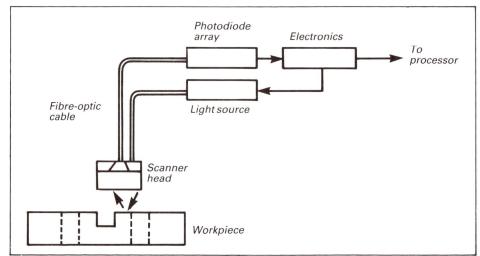

Fig. 1   Fibre-optic 'vision' sensor concept

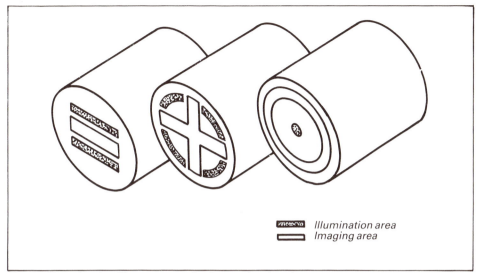

Fig. 2   Terminating fibre geometries in sensing head

This approach to vision sensing offers several advantages:

● It allows physical separation of the electronics and optics. As a result, a small scanner head can be integrated flexibly with the robot's gripper, thus allowing it to 'see' in hard to reach areas without getting in the way.

● The scanner head and the fibre-optic cable are lightweight, thereby minimising any degradation in the robot's performance. In addition, the fibre-optic cable provides excellent immunity to electrical noise, resulting in increased reliability.

● There is the potential for geometric preprocessing by arranging the optical fibres in a suitable pattern in the sensing head, while the other end of the aligned fibres is matched to the photodiode array. An example of various fibre geometries for imaging as well as illumination in a sensing head is shown in Fig. 2. The linear, cross and circle patterns can very quickly provide area, centre location and angle information, respectively.

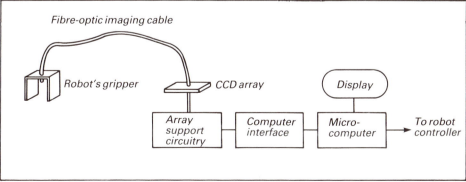

Fig. 3   Fibre-optic eye-in-hand system

This simple preprocessing can in turn reduce the amount of processing needed in the computer, thus improving the closed loop response. An experiment using a cross-pattern for hole centre location was performed which showed linear response with extremely good sensitivity[6].

Fibre-optic eye-in-hand system

A fibre-optic eye-in-hand system is shown schematically in Fig. 3. It consists of a fibre-optic imaging cable, CCD array, array support circuitry/computer interface and microcomputer. These components are described as follows.

Imaging cable

The imaging cable consists of a bundle of fibres (typically $10\mu m$ dia.) housed in a flexible sheathing such as rubber or plastic. As an example, a 100×100mm square cable with a packing factor of 0.7 will contain over 40,000 such fibres for image transmission. The fabrication of flexible imaging bundles involves precise winding of the optical fibres on a highly polished and uniform cylinder whose circumference determines the length of the imaging bundle. The aligned fibres can be wound directly from the fibre drawing furnace or from a separate spool of non-aligned fibres. During drawing and winding, the fibres are usually soaked in a high-viscosity binding fluid. When the winding is completed, the layers of fibre are secured, cut and removed from the drum. The resulting layer has two ends which are precisely aligned. These ends are bound to retain the orderly arrangement, while the rest of the length is left loose by removing the binding fluid, thereby rendering the device flexible.

CCD array

The imaging cable is coupled to a CCD array in such a way that the light coming through the fibres falls onto the diode array without any angular dispersion. This is accomplished by either mechanically coupling the cable end to the array or using optics to focus the image at the cable end onto the diode array. It should be noted that the size of the fibre in relation to the diode size will affect the overall system performance in terms of edge definition. The diode array converts incident light into electric charge, which is integrated and stored in a shift register. In addition, MOS transistor switches are used to multiplex photodiode outputs to a single serial output line via a shift register. The diode array, shift register and multiplexing switches are integrated into a single silicon chip.

CCD arrays offer several advantages over vidicon tubes:

- Their lack of retention of previous images allows completely new information to be read out at each scan of the array.
- Excellent positional accuracy of each diode (0.1mm or better) makes them useful for accurate measurements.
- They are most sensitive in the visible and near infrared range, which includes the wavelengths of light emitted by many inexpensive sources.

Array support circuitry computer interface

Support circuitry is needed to provide control for clocking the array and processing its output signal via a sample-and-hold circuit. A continuous analogue signal is generated corresponding to outputs from the diodes. A typical maximum signal is about 3V at saturation. This signal is converted into digital form for computer processing. For binary images a threshold is used to differentiate between light and dark regions. For grey-scale images an appropriate A/D converter can be used for the required accuracy.

It is desirable to preprocess the data to reduce computer processing requirements. One approach is to store only the position of every light-to-dark or dark-to-light transition for each scan line instead of each array element output. This is generally known as 'compressed line' boundary representation. In addition, the total number of transitions in a given scan line is another useful number providing edge information. For continuous acquisition of the data, at least two buffer memories are needed. While the compressed data from one scan is being loaded into one memory, data from the previous scan can be accessed by the CPU. After these operations are completed for one scan, the two memories are functionally swapped, which allows the new data to be available to the CPU without any interruption. Once the data is in the computer it can be processed and displayed as required.

Experimental results

To prove the feasibility of the concept, a system was built using the design approach described in the previous section. The system consisted of a 4ft (1.22m) coherent fibre-optic cable with a cross-section of $150 \times 2mm$ using $10\mu m$-diameter glass fibres. On one end it has a 26mm screw-on type lens, while at the other end it is mechanically coupled to a 128-diode linear array with 1mm spacing between diodes. The cable and the diode array are shown in Fig. 4. A threshold was used to convert analogue video into binary data.

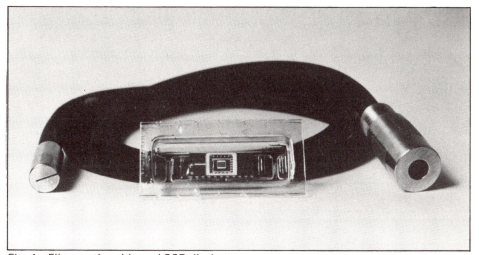

Fig. 4   Fibre-optic cable and CCD diode array

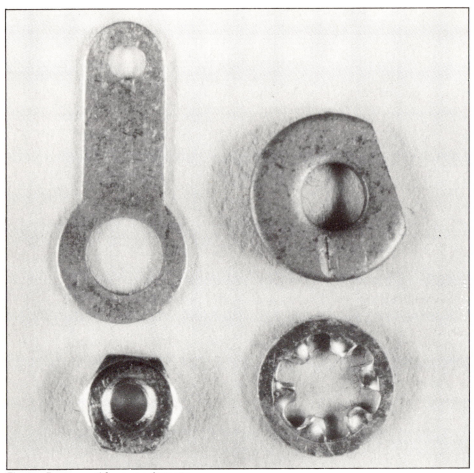

Fig. 5   Parts used for experiments

The output was interfaced to a microcomputer with ability to control the scan, acquire the data and store it in a file. A 512-line resolution graphic display was used to examine scanned images. The scanner head was mounted in a gripper on a Cartesian robot which was developed in-house and is described elsewhere[7]. Parts for experiments were selected having various geometric features (edge, circle, curve, protrusion, etc.) as shown in Fig. 5. The sizes ranged from 0.1 to 0.5in. (2.5 to 12.7mm).

It is important to recognise that lens and image sensor selections are of key importance for a given application. One needs to consider field of view, acceptable working distance, object accuracy and object resolution. From these, one can determine the required array length, resolution and lens focal length[8]. In our experiments, a field of view 0.5in. (12.7mm) wide was chosen, which resulted in an object resolution of 0.004in. (0.1mm) when using a 128 diode linear array on 0.001in. (0.025mm) centres. For a 26mm lens, a working distance of about 5.5in. (140mm) was obtained. With this arrangement, experiments were performed under backlighting and front-lighting as described below.

Fig. 6   Back light test set-up

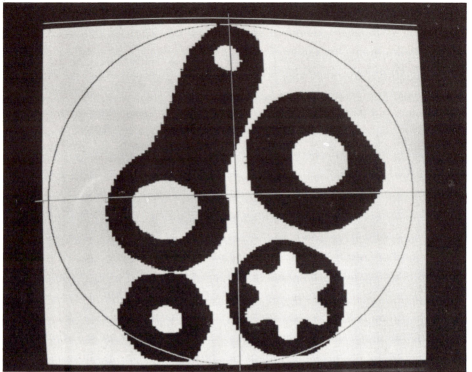

Fig. 7   Scanned image under backlighting

Backlighting

A light-table was used with back illumination from an incandescent source, which provided a relatively flicker-free signal in comparison with that obtained from a fluorescent source. The scanning head was mounted on the side of the gripper as shown in Fig. 6 for clarity of presentation. However, it can be easily integrated into the robot's gripper. The parts shown in Fig. 5 were scanned by moving the robot at a constant speed of 1in./s (2.54cm/s). The resulting image formed during scanning is shown in Fig. 7. It appears that excellent edge definitions were obtained when one considers that only a 128-diode linear array was used.

Frontlighting

A similar experiment was performed under frontlighting. The same parts were placed on a black background (for contrast) and were illuminated using a tungsten source with light transmitted by fibre-optic cables as shown in Fig. 8. The parts were similarly scanned, resulting in an image as shown in Fig. 9. It required some adjustment in light placement and threshold to get reasonable images at the same time for all parts of different thicknesses. The lack of edge sharpness is due to shadow effects (especially in thick parts such as the nut) and varying angle of reflectance at the edges. Further improvements can be made by optimising the illumination and using a larger array size.

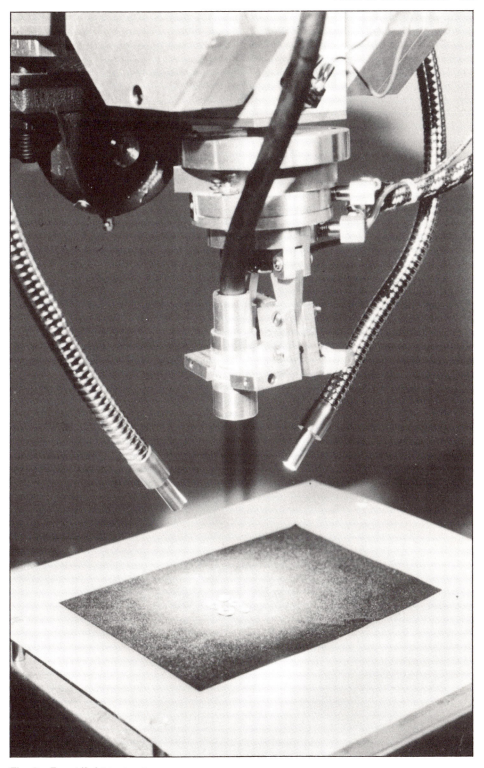

Fig. 8   Front light test set-up

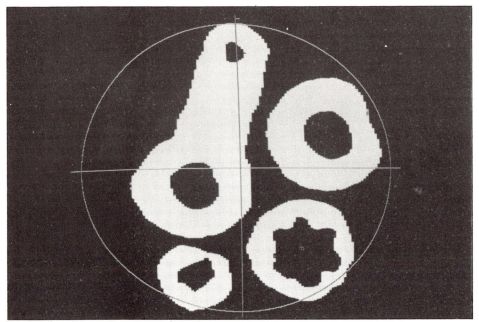

Fig. 9   Scanned image under frontlighting

Processing for sensory control

Besides physical integration of the vision sensor with the robot's gripper, the key to visual servo control is tight coupling of information from the sensor to the robot's controller for servoing[9]. A robot visual servo control is shown schematically in Fig. 10, where the vision processor is in the feedback loop. The objective is to use the time-varying image features to guide the arm trajectory in real-time. In our implementation, separate LSI-11 micro-computers, communicating over a 16-bit parallel bus, are used for robot control and vision processing. The two processors work in parallel to reduce the overall servo loop processing time. The necessary software interface to accommodate sensory data was developed in-house[7].

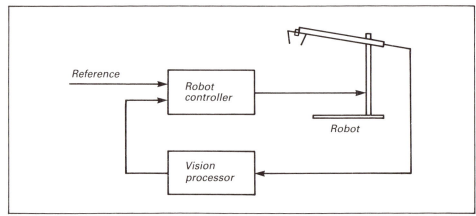

Fig. 10   Robot visual servo control

To reduce the computation time, one can take advantage of several properties of the system described in this paper. Since one is scanning the object line by line and there is continuity of image features, one needs only to examine changes and compare them against certain criteria to interpret the significance of these changes. Since there is a large amount of redundant information in the image, a large fraction of these changes can be discarded immediately since they do not contribute to the feature information one is seeking. This approach also reduces data storage needs, since at any given time one is storing only the current data line and perhaps a few items from a previous scan. It appears that for many situations one may extract the necessary features without even completing the scan.

As mentioned earlier, another advantage offered by optical fibres is the ability to shape them in suitable patterns (Fig. 2) according to the object features that need to be detected. This idea can be expanded further when several fibre-optic scanning heads can be strategically positioned in a gripper while interfaced to a single CCD array. At any given time during the motion of the gripper, appropriate diodes on the array are interrogated for output. This is analogous to having several cameras but sharing the electronics under common control. These 'cameras' are spatially independent yet tightly coupled with respect to the feedback information for servo control.

Concluding remarks

Based on the experiments described in this paper it can be concluded that good images can be obtained under back and frontlighting conditions at relatively high speeds utilising a coherent fibre-optic bundle interfaced to a CCD array. It has also been demonstrated that a small scanning head can be mounted flexibly on the robot's gripper, thus allowing fabrication of an eye-in-hand. In addition, this approach allows for significant simplification in processing for real-time gripper servo control.

Acknowledgements

The authors would like to thank G. Katzan for integrating various elements of the system, N. Herbert for developing the test software and C. Kot of Northwestern University for fabricating the fibre-optic cable.

References

[1] Loughlin, C. and Hudson, E. 1983. Eye-in-hand robot vision scores over fixed camera. *Sensor Review*, 3(1): 23-26.
[2] Makhlin, A.G. et al. 1982. Vision controlled assembly by a multiple manipulator robot. In, *Proc. 2nd Int. Conf. on Robot Vision and Sensory Controls*, Stuttgart, pp. 83-92. IFS (Publications) Ltd, Bedford, UK.
[3] Nagel, R.N. et al. 1980-81. Experiments in parts acquisition using robot vision. *Robotics Today*, Winter 1980-81: 30.
[4] Pugh, A. et al. 1983. A research program in sensor guided assembly. In, *13th Int. Symp. on Industrial Robots*. SME, Dearborn, MI, USA.
[5] Epstein, M. 1980. Endoscopy: Developments in optical instrumentation. *Science*, October 1980: 200.

[6] Agrawal, A. et al. 1983. Machine vision based on fiber optic imaging. *Electronic Imaging*, October 1983.

[7] Masory, O. and Jakopac, D. 1983. An experimental robot based assembly cell. In, *Proc. 1983 ASME Conf.*, Chicago, USA, p. 15. ASME, New York.

[8] Hopwood, R.K. 1980. Design considerations for a solid-state image sensing system. *Proc. SPIE*, 230: 72.

[9] Sanderson, A.C. and Weiss, L.E. 1982. Image based visual servo control of robots. *Proc. SPIE*, 360: 164.

OPTICAL ALIGNMENT OF DUAL-IN-LINE COMPONENTS FOR ASSEMBLY

P. A. Fehrenbach
GEC Research Laboratories, UK

In complex systems for assembling small components, errors in the various parts can build up to a significant but unknown relative error between two components to be mated. If this error cannot be removed then assembly may not be possible. A system for aligning dual-in-line circuit packages over pre-drilled holes in a circuit board, which overcomes this difficulty, is presented. The system comprises an internally illuminated $XY\theta$ table on which the board is mounted, a gripper for holding the packages with optical fibre sensors for detecting the correct alignment, associated interface hardware and control, and communications software. The system forms part of a much larger flexible assembly system for circuit board components.

The insertion of dual-in-line packages (DIPs) into printed circuit boards requires that the pins of the packages and the holes in the boards are aligned accurately to avoid damage to the package pins. The accuracy required depends upon the size of the holes relative to the cross-section of a pin; a standard-sized pin needs more accurate alignment over a 0.5mm diameter hole than over a 1.0mm hole. The ability of a manipulator to position a component with the required accuracy at a point in space to which it has never been before, but which it has calculated in its own coordinate system following calibration at other known calibration positions, is limited. The resolution of the manipulator is also often insufficient to make the necessary corrections for any error in initial alignment as detected by sensory feedback. A way of overcoming this problem is to use a second manipulator with greater resolution but a smaller range of movement than the first. This second manipulator is used to make adjustments to the position of the circuit board while the DIP is maintained at a constant position by the first manipulator as a result of its first and best attempt at alignment. This approach of using a coarse manipulator with a large range of movement and a fine manipulator with a much smaller range is sometimes called the 'right arm and left hand' approach because of the way it parallels the technique used in manual assembly of picking up objects with the right hand, moving them to the centre of the work area and assembling them using two hands.

In the system described here the movements of the fine manipulator are determined by the inputs to optical sensors on the end-effector supported by the coarse manipulator. These 'see' where the holes in the circuit board are and enable the board to be moved to position the holes directly under the pins of the package.

System overview

The alignment operation is one of several needed to assemble components on a circuit board correctly and constitutes only a small part of the overall assembly system. The system in total comprises a coarse manipulator and end-effector, component slide feeders, a fine manipulator, interfacing and processing hardware and control and communications software. The system has been designed and developed as a research tool to illustrate one application of flexible assembly systems. The software runs on a multi-processor system and is divided between four processors. Each of the processors runs several program modules or processes and the alignment controller is one process running on a single processor. The software has been designed so that inter-process communication is processor indepen-dent. The 'Align DIP' function of the alignment controller is requested by the insertion controller at the appropriate point in the assembly sequence for each DIP. The request includes information about how many pins the DIP has and its position and orientation relative to a fixed datum in the plane of the circuit board. After attempting to align the package with a set of holes in the circuit board, the alignment controller responds with either a success or failure signal to the insertion controller. Although there are many possible reasons for failure, the response contains no additional information because the insertion controller provides only one type of failure recovery, by rejecting the DIP to a bin.

Fine manipulator – the $XY\theta$ table

Mechanics

The fine manipulator is an $XY\theta$ table. The table is driven by three dc servo motors and illuminated from underneath by tungsten filament lamps. The lamps are supplied from a dc power source to give a constant level of light and are housed beneath a diffuser panel. A cross-section through the table which shows this arrangement is provided in Fig. 1.

The three motors are arranged as shown in Fig. 2, with two on the same side of the table and the third on an adjoining side. Each motor has a cam attached to the end of its shaft. Towards one end each cam has a ball-bearing assembly which can run in a slot in the bottom surface of the removable table top. The other end of each cam has a vertical flag which passes between an infrared emitter and a detector as the motor turns. This arrangement is used to datum the table following power up.

The offset of the ball-bearing assembly from the motor shaft on each motor means that moving one of the motors causes its ball-bearing assembly to move in an arc. This movement is resolved into a component parallel to the slot in the top of the table and one normal to it. The parallel component

Fig. 1   Sectional elevation of XYθ table (motor assemblies not shown)

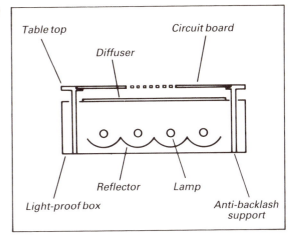

Table top

Circuit board

Diffuser

Reflector

Lamp

Light-proof box

Anti-backlash support

is taken up by the assembly sliding in the slot and the normal component causes the table top to move. Fig. 2 shows the directions of the axes used when describing table moves and numbers the motors from 1 to 3. Moving motor 3 only moves the table parallel to the X-axis, while moving motors 1 and 2 in opposite directions by equal amounts moves the table parallel to the Y-axis. If motors 1 and 2 are moved in the same direction or moved by unequal amounts then the table rotates about an axis normal to the plane of the table top. The motor movements all interact with one another so that moving one will cause another, which does not move, to slide in its slot. Once a rotation has taken place the slots in which the bearing assemblies slide are not parallel to the X- and Y-axes and motion parallel to these axes is then more complicated than that described above.

The range of movement of each of the motors is very restricted. Partly this is caused by interaction with the other motors so that some positions are only accessible to a motor when the others are in certain regions of their own

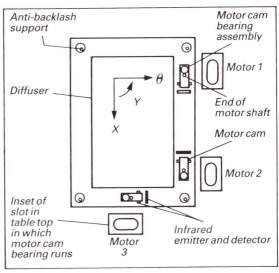

Anti-backlash support

Motor cam bearing assembly

Motor 1

Diffuser

End of motor shaft

Motor cam

Motor 2

Inset of slot in table top in which motor cam bearing runs

Motor 3

Infrared emitter and detector

Fig. 2   Plan of XYθ table with table top removed

ranges. The most restricted range is that of motor 2, which can move the table top only about 4.5mm. This is quite sufficient for the alignment of DIPs. The restricted range of rotation available simplifies the control calculations by always allowing the small angle approximations for θ less than 5° to be made.

Control

Position information for each motor is provided by an incremental encoder and a 16-bit up/down counter. The counter registers 5838 counts per revolution of the motor output shaft. At power up the counters are all automatically cleared and the software prevents any other function being requested until the motors have been datumed. This is done for each motor in turn by driving it back and forth so that both edges of the flag on the motor cam interrupt the infrared beam. The counter readings when the detected signal crosses a threshold are recorded and the mean value calculated. This counter reading represents the position where the flag is central in the beam and blocking it completely. The motor is moved to this position and the counter cleared. This is then the datum position and the full range of movement is approximately symmetrical about it.

The motors are moved by applying a voltage V according to the formula:

$$V = k_1 e - k_2(e - e_p) + I \qquad (1)$$

$$
\begin{aligned}
I &= k_3 e + I_p, & \text{when on} \\
&= 0, & \text{when off}
\end{aligned}
\qquad (2)
$$

where e is the error between the required and actual position, e_p is the error value from the previous voltage calculation, I is an integrator term which is not always present, I_p is an integrator value from the previous voltage calculation, and k_1, k_2, k_3 are constants.

This constitutes a PID controller in which the integral term is turned on and off at certain times as explained below. The controller also incorporates a programmable current limit for each motor which limits the torque that it can apply to prevent mechanical damage or electrical overheating.

The motor servo controller must be able to perform two different but related tasks: it must be able to move the motors to some new specified position as quickly as possible and it must be able to hold the motors stationary at a given position against the force from the anti-backlash supports. The system is biased in a highly non-linear fashion by the anti-backlash supports and the interactions of the motors, so the servos are very often active when the table is left in one position for a period of time.

When a move is requested the request specifies the destination position and current limit for each motor. In order to kick the motors away from rest quickly, the current limit is set to the maximum possible value for the first four update periods of each motor. It is then set to the requested value. At the beginning of a move the integrator term (Eqn. (2)) is turned off because a large voltage is applied by the proportional term in the controller and if the integrator was included it would soon integrate up to a large value which would swamp the other terms. The integrator for any motor is turned on when that motor first comes to rest after the start of a move and it remains on until the move is complete. By this time the positional error should be small and the integrator will not reach a very large value. A motor is deemed to have reached its correct destination when either one of the following conditions is satisfied:

- The motor has stopped within the deadband of one count on either side of the requested destination.
- The motor has moved only a small distance during the previous four servo update periods and is presently within the deadband about the requested destination.

The move is considered to have been completed when all the motors have satisfied one or other of these conditions or the maximum time allowed for a move, one second, has expired. The move function then responds with a message to the requester indicating either success or failure and sending the actual final positions attained by the motors as parameters.

During a move the controller cycles round the servo loop as fast as the processor will allow and no other functions have access to the processor. The time between voltage updates is thus optimal and the move is completed as quickly as possible. When the motors are being servoed on a constant position, however, the servo is only updated every 10ms. This frees the processor to service other requests when they arise but keeps the table in the correct position. It is to this type of servoing that the table automatically changes when it has completed a move, whether it succeeded or not.

Optical sensing

Hardware

The hardware for the optical sensing used during alignment of a DIP is all mounted on the end-effector of the coarse manipulator. The end-effector has sixteen fingers in two opposing rows of eight for handling DIPs with up to sixteen pins. A 0.010in. diameter plastic optical fibre is cemented in a groove in the inner face of each finger and passes the whole length of the finger, projecting a small distance beyond the end. This end of each fibre is cleaved and the other end embedded in a ferrule and connected to a PIN diode detector. The detectors are mounted on top of the end-effector close to their respective amplifiers which produce ±10V outputs for subsequent processing. This arrangement allows the fibres to be located immediately next to the pins of a DIP being aligned, while at the same time affording some protection against accidental damage. With such a small fibre in this position, when the centre of the fibre is over the centre of a hole in the circuit board the adjacent pin is also sufficiently central over the hole to allow it to be inserted. The end-effector is shown in Fig. 3.

Interfacing between the sensor hardware on the end-effector and the control processor is achieved using a sixteen channel analogue-to-digital converter. This is a 12-bit device which provides 4096 levels of output.

Fig. 3   The end-effector

Software

The system handles packages with from two to sixteen pins and always grips a package with a given number of pins using the same fingers. The software knows how many pins the DIP being aligned has and, therefore, knows which optical fibres have pins adjacent and which do not. When scanning all the pins of a package this enables the program to look at only those fibres it needs to instead of reading all sixteen inputs. At another stage in the program it is necessary to read the input from one fibre only, and this can be done with no difficulty. Once the sensor data has been read into the processor in this way it is used to control the alignment of a DIP as described below.

Alignment strategy and tactics

The aim of alignment is to move the $XY\theta$ table to maximise the light input at all the optical fibres which correspond to pins of the DIP. Ideally all these maxima would occur at the same position, but in practice this is not usually the case and a lesser criterion is adopted. The strategy employed is to maximise the light at any one fibre first of all and then rotate about that fibre so that all the others are maximised simultaneously. The DIP can then be inserted in the board.

The tactics employed to achieve this are governed by how a fibre responds to a hole in the board and to the absence of a hole. Typical responses when a fibre is used to scan an area of circuit board are shown in Fig. 4. These show

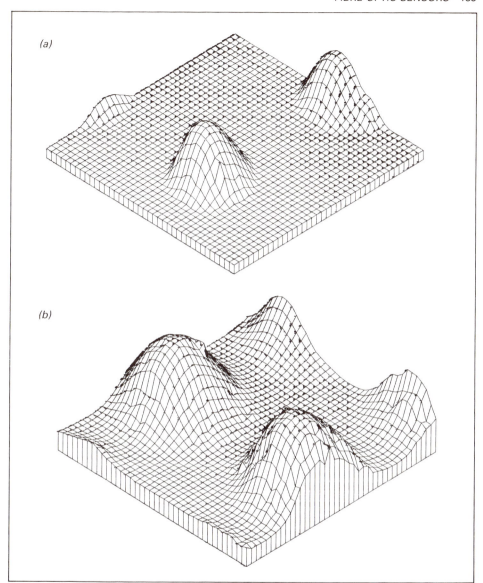

(a)

(b)

Fig. 4   Response of an optical sensor to holes in a circuit board. Plotted readings are on a 0.1mm grid over an area 4.0mm square: (a) 0.5mm diameter holes 2.45mm apart, (b) 1.0mm diameter holes 2.45mm apart

how a fibre responds to holes of different sizes and how it inteprets the blank area of board between holes. The plots of Fig. 4 show a clear contrast between the holes and the spaces between them. The sensors rely on transmitted light, however, and the contrast is dependent on the opacity of the circuit board. If a transparent board is used the sensors cannot distinguish between a hole and a blank area of board. They can detect tracks and pads but the system is not designed to use these for alignment. The standard height of the fibre end above the surface of the board when being used as a light sensor is 1mm.

First of all the fibres have to find a hole. This is indicated by detecting a variation in the light input at one of them. It may be that a fibre is over a hole initially, in which case it is only necessary to confirm this, or it may be necessary to conduct an orderly search over a predetermined area to try and find a hole, if all the fibres begin in the regions between holes where the response is flat. It is assumed that the first hole detected for each fibre is the correct hole for insertion. This is true if the coarse manipulator has positioned the DIP within half an inter-hole distance of the correct holes. Ideally the DIP would be positioned exactly over the centres of the holes to start with, and the coarse manipulator will have done this as accurately as it can. There should, therefore, be a hole very close to each fibre to start with, and the probability of finding a hole will decrease with distance from the initial position. A search is needed which begins close to the initial position of the fibres above the board and only searches further away if it has failed to find a hole at a closer distance. The search must cover a square area of side equal to the inter-hole distance (2.54mm), centred on the initial position and with sides aligned with the table X- and Y-axes. The search must be sufficiently fine to detect the smallest hole into which a pin can be reasonably inserted anywhere in that area. Such a search is implemented by the pattern shown in Fig. 5. The search starts in the centre and spirals outwards reading all the fibres at 0.2mm intervals along the path until a hole is detected. A square spiral was chosen as opposed to any other pattern, e.g. circular spiral, because it covers the whole of the search area, including the corners, without going outside the boundaries and because all moves are parallel to the table axes. The latter reason means that calculation of the next position is simplified and also that it is not necessary to move all the motors for any one step along the search path. This makes the calculation and movement fast.

Once a hole has been detected then hill climbing is used to find the centre; the centre is the brightest part of the hole. First, the light entering the fibre which detected a hole is maximised by moving the table in the X-direction only. While holding the table at this X-displacement, the light is then further maximised by movement in the Y-direction only. A check on the X-direction maximum follows and the hill climbing is then complete. The table is rotated about the maximised hole until the other fibres record maximum inputs. Alignment is then finished.

Realisation of alignment theory – The software

Throughout the alignment of a DIP, how the $XY\theta$ table moves is determined by the software on the basis of the two latest sets of readings from the optical sensors, in accordance with the tactics described above. The table is moved in small steps of differing size, according to the part of the program being executed, and after each step appropriate fibres are read. When a DIP has been positioned by the coarse manipulator and the optical fibres are read for the first time, there are two classes of possible outcome: either all the inputs return the same value or at least one input is different from the rest. Very little can be determined from a single set of readings, but of these two classes of result the latter, showing a spatial variation of light level over the package area, suggests more strongly that at least one hole has been found. For the results to be useful the table is moved a small, arbitrary distance and the fibres read again. From these two sets of readings it is possible to tell whether a hole has been found or not. If not, then a search is initiated. If a hole has been found then hill climbing begins immediately.

When searching, each new fibre reading is compared with the previous reading at the same fibre. If there is a significant difference between any of the pairs of readings then a hole has been detected and the search is ended. A software filter is used when comparing fibre readings to eliminate noise and ensure that small, local maxima in the light level are not mistaken for holes. The same filter is used during hill climbing. Hill climbing involves reading one fibre only and comparing the two latest readings. Once it has been established in which sense, positive or negative, the table should move parallel to the allotted axis, the table continues to move in the same direction until a significant decrease in the light level from one fibre reading to the next is detected. The table moves in increments of 0.1mm. When a decrease in light is observed the table reverses its direction of motion and steps in 0.02mm increments for as long as the light levels read continue to increase significantly. This implementation is used for both the X and Y parts of the hill climb. The final X correction is similar but entirely at a step size of 0.02mm.

The searching and hill climbing operations are reliant solely on relative light levels. This means that the background level of illumination is not critical to the operation of the system. However, this may lead to the mistaking of some other, local light maximum for a hole, though the filter should prevent this, and will produce a spurious result if the lamps beneath the circuit board are turned off or if there are certain types of hardware fault, such as a broken optical fibre. As a first check to prevent this, the light input at the maximised fibre following hill climbing is compared with an absolute threshold level. The light level at this fibre should be very high if the lamps are on and there are no faults. The threshold is chosen to determine whether or not this is the case, and if not, to abandon the alignment altogether on the grounds that there is a hardware or other non-recoverable fault. If the light level is above the threshold then alignment proceeds without interruption.

Rotation about the maximised fibre is similar to hill climbing but in one dimension only. As in hill climbing, only one fibre is monitored and this is chosen to be the brighter corner fibre on the other side of the DIP to the maximised fibre. The table is incrementally rotated in a positive sense about the maximised fibre. If the light input at the monitored corner fibre decreases, then the sense of rotation is changed to negative and the table rotates back to its previous position. Incremental rotation and corner light monitoring continue repeatedly in the chosen rotation sense until the light input at the corner decreases. The table then rotates back incrementally to its previous position and the alignment movements are complete. Rotation only takes place at all if the light level at the brighter corner on the opposite side of the DIP is significantly darker than that at the previously maximised fibre. If this is not the case, then the rotation sequence is bypassed. This condition is checked again when rotation has finished, and if the corner is still significantly darker then alignment is deemed to have failed.

When all the alignment movements of the table have been completed, then a final check is run on all the fibres corresponding to DIP pins before the software responds that alignment has finished. The light input at each fibre is compared with a threshold level designed to ensure that each fibre is reasonably illuminated. If any of the fibre inputs falls below this level then alignment has failed. This detects blocked or missing holes in the circuit board and prevents insertion taking place which may damage the DIP or the board. Another check is run at many different places in the alignment sequence on the light input to the currently monitored fibre. If it has saturated the hardware because of a very high light input, then it is pointless trying to increase the input further because it could not be differentiated from the present state. The effect of such a condition depends on the point at which it occurs in the sequence of movements, but generally it results in the program abandoning the current stage and proceeding to the next stage of alignment.

The software responds to the requester either when one of the many tests fails, which results in a failure response, or when all the table moves have been successfully completed and the final checks all passed with no failures.

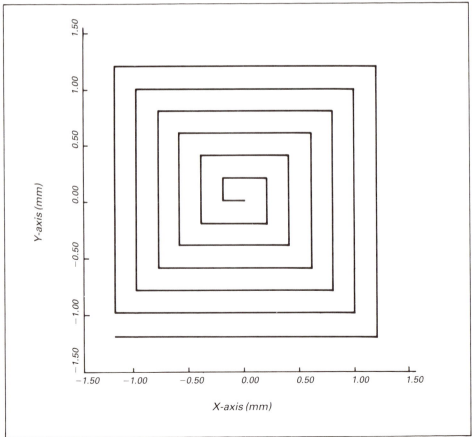

Fig. 5   Whole of search path, starting from (0, 0), followed by XYθ table if no hole is found

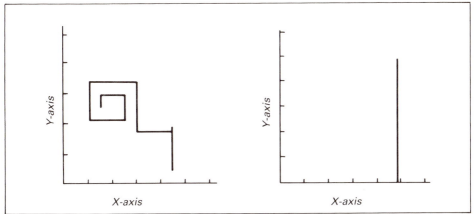

Fig. 6 *Typical paths followed by XYθ table during alignment. Movement always starts from (0,0): (a) search and hill climb for 0.5mm diameter hole, (b) hill climb only for 1.0mm diameter hole*

This latter case produces a success response. There are no parameters associated with the response, whether success or failure, and in the case of failure there is only one recovery strategy and the DIP is rejected to a bin. The reasoning behind this is that, unless an insertion can take place perfectly after proper alignment, it is not worth attempting.

System performance

Experiments with the whole printed circuit board assembly system have shown that the alignment is capable of playing its role adequately. The time taken to align depends on the diameter of the holes in the board and on the

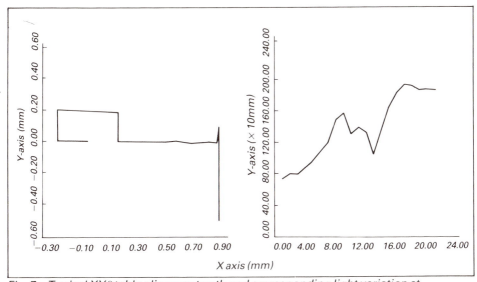

Fig. 7 *Typical XYθ table alignment path and corresponding light variation at monitored optical fibre during hill climbing section for 0.5mm diameter hole: (a) alignment path starting from (0,0), (b) light input at optical fibre monitored during hill climbing against number of table moves since hill climbing began*

initial degree of misalignment. Typical times taken for successful alignment are 3.0s for 1.0mm diameter holes or 3.5s for 0.5mm diameter holes. Times as short as 1.0s and as long as 5.5s have been recorded. The exact path taken by the fine manipulator to reach its final position varies greatly depending on the initial misalignment; some typical paths are shown in Fig. 6. In the majority of cases no search is needed and the table is able to begin hill climbing immediately. The way the light input at the monitored fibre varies during hill climbing can be seen in Fig. 7. This shows a typical alignment path plotted in *XY* coordinates and the light input at the monitored fibre during hill climbing plotted against table movement, so that each new point signifies another movement by the table. The maximisation of the light, first in one direction and then in a perpendicular direction, can clearly be seen, together with a final check in the first direction.

Concluding remarks

The alignment system described has been shown to work in a well-behaved environment and to position the holes in a printed circuit board correctly for insertion of DIPs. Some parts of the system are rather intolerant to noise and require delicate tuning to function reliably, but the system has achieved its main aims by demonstrating that the alignment algorithms work properly and integrate with the rest of the assembly system.

DYNAMIC SENSING FOR ROBOTS – AN ANALYSIS AND IMPLEMENTATION

G. Beni, S. Hackwood and L. Rin
University of California, USA

A systematics of robot sensor design has been initiated by formulating the general problems and addressing the specific question of how to arrange sensing elements. A fibre optic sensor for the fingers of an Intelledex 605 robot has thus been constructed. The sensor consists of a linear lens array of eight elements attached to the edges of the robot fingers. The elements are composed of parallel, equally spaced, collimated light beams that pass from finger to finger. Using this sensor the location, orientation and shape of objects within a robot hand can be determined. Using dynamic sensing three-dimensional images with a 0.5mm resolution can be reconstructed.

The development of intelligent robots requires research on new sensors and the software necessary for their use. The most studied types of sensors are visual[1], proximity[2], and tactile[3]. There are also sensors that cannot be easily classified into these groups. One such sensor is the recently proposed 'dynamic' optical sensor[4], which is used for help in acquisition as well as in inspection of objects.

The design of this sensor follows a new criterion, an essential feature being its dynamic operation. For most sensors proposed and/or implemented so far, the sensing field is constant in time, as, for example, for a fixed camera or tactile pad. Exceptions are cameras mounted on robot hands[5] and fingertip sensors for the three-fingered Stanford/JPL hand[6]. In both these cases, robot-manipulation is used to increase the sensory information about the object investigated, i.e. the sensing is a dynamic process. Similar to these cases, our noncontact inspection sensor takes advantage of the robot dynamics.

It is not always true that dynamic sensing is superior to static sensing. In general, the choice between static and dynamic sensing arises from the competition between the cost of adding sensing elements versus the cost of moving the sensing elements. (Cost here is intended in a general sense, i.e. not only financial but also functional.) In robotics, dynamic sensing may be *a priori* favoured since the cost of introducing mobility is negligible if one takes advantage of the robot motion itself.

Clearly, if a reduction in the number of sensors is not outweighed by the cost of introducing mobility, a more efficient design is obtained. This assumes that performance of the sensory system is not diminished. Thus a careful balance must be achieved between:

- The number of sensing elements.
- Sensing element speed of motion.
- Response time.
- Processing time.

More precisely, we have formulated[5] the basic problem of sensor arrangement and design as follows: *For a given sensor (characterised by its own response and processing times), what is the relationship between speed and number of sensors so that the combined response plus processing time for the same sensing field remains constant?*

We have shown that a small number of sensors (linearly) arranged on a robot can provide a substantial amount of object shape information, comparable to that obtained from a much larger tactile pad. Thus we have designed a linear dynamic optical sensor to be mounted on the finger of an Intelledex 605 robot. With minor modification it could be used on other robots. The only requirement is that the fingers of the robot gripper maintain a parallel alignment during opening and closing.

Sensor design

The dynamic optical sensor consists of a linear lens array mounted near the edge of one finger of the robot hand. This produces eight parallel, collimated light beams between the fingers, which are monitored by a linear detector array mounted on the opposite finger. The sensor detects the presence of objects between the fingers when one or more of the light beams are interrupted.

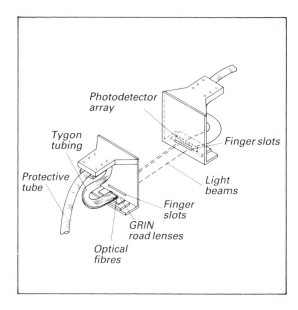

Fig. 1   Detailed view of sensor on fingers

Fig. 1 is a view of the sensor as mounted on the fingers of the robot. The fingers are removable from the main body of the gripper and therefore can be designed for a specific task. In this case, the robot fingers are two inches square and made of aluminium. Optical fibres are brought up to the sensor array through a robust protective tube. Each fibre is further protected by a thinner piece of tubing which runs from the end of the larger tube to the sensor itself. The main body of the sensor consists of an array of eight graduated refractive index lenses which collimate the light brought to them by the optical fibres. Beams of light from a He-Ne laser source travel across the space between the fingers and are detected with a silicon photodetector array. The fingers are designed so that the sensor is mounted close to an edge. This does not hinder object handling nor the placement of additional sensors in the finger pads. The sensor attaches to the back of the finger and is protected from accidental collision while the robot is in motion. Small slots are cut through each finger to allow the beams to pass between them.

As shown in Fig. 2, in normal ambient lighting with a finger spacing of 3cm the voltage output V is ~50mV. The voltage difference $\triangle$V between an uninterrupted and a blocked beam is 30mV. In the dark, V is ~120mV and $\triangle$V is >80mV (see Fig. 2(c)).

With a finger separation of 5cm, the strength of the signal was only about 10mV lower (~5%) (see Fig. 2(b,d)). Thus opening and closing the fingers will not affect the sensor's ability to detect the presence of an object between the fingers. The signal-to-noise ratio in a well-lit area was greatly increased by placing an interference filter tuned to the wavelength of the laser (633nm) directly in front of the photodetectors.

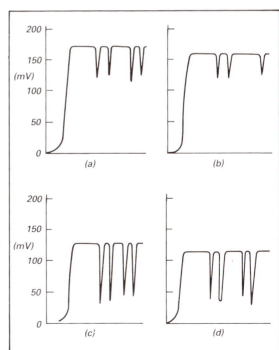

Fig. 2 Data showing the detection of an object between the fingers, and the change in intensity of the signal with finger separation. (a) and (c) show the detection of a presence with a finger separation of 3cm, while (b) and (d) show that at a separation of 5cm. For (a) and (b), the room is well lit; (c) and (d) were recorded in a darkened room

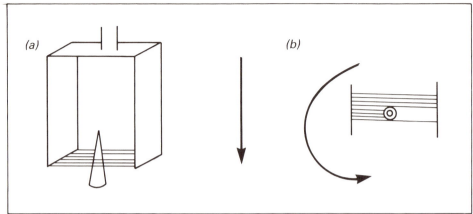

Fig. 3   Two possible sensor configurations and two different scanning methods. (a,b) one possible arrangement of light beams: a linear array. (a) Linear scan: orientate the light beams horizontally and move the sensor vertically, or orientate the beams vertically and move the sensor horizontally; (b) cross-section scan: orientate the light beams horizontally and rotate the sensor in that plane

In spite of its simplicity, the optical sensor can be used to extract quite detailed information. Typically the robot scans in two ways: linearly, in the plane orthogonal to that of the light beams, or cross-sectionally, in the same plane as the beams, depending on how the sensor is orientated and moved. When the line of light beams is positioned horizontally and moved vertically in a straight line along the Z axis, as in Fig. 3(a), a two-dimensional projection of the object is detected. Orientating the light beams vertically and moving horizontally in a straight line along the Y axis accomplishes the same effect. When the beams are positioned horizontally and rotated around the object, as in Fig. 3(b), a two-dimensional cross section of the object at the height Z of the beams is detected. A substantial amount of information can be gathered with the simple configuration of Fig. 3(a,b) as follows.

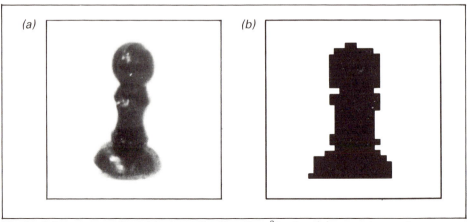

Fig. 4   A linear scan of an object smaller than 2cm^2 in size: (a) pawn chess-piece; (b) linear scan of pawn

The image of the object, whether a projection or a cross-section, is stored in a two-dimensional array. The area of each individual element, or pixel, is the square of the width of the beam. If the image is a cross-section, its Z-coordinate is stored also. For simplicity, we have restricted the discussion to purely vertical and horizontal (YZ- and XY-) planes, although slanted planes can be dealt with similarly.

Forming an image from a linear scan is quite straightforward. As the sensor relays data, it is put directly into the array, and eventually a projection of the object emerges. Two examples are shown in Fig. 4.

Forming an image from a cross-section scan, however, is slightly more complicated. The sensor is rotated around the object and data is taken at different angles. Taking data at six different angles usually gives sufficient amount of detail to distinguish the object, without requiring an exorbitant amount of processing and scanning time. Thus we have chosen to rotate the light beams clockwise to the angles 0°, 90°, 30°, 120°, 60° and 150°.

The starting image is always a totally filled area, and as more data is gathered at different angles, more is discovered about the boundaries of the object. Definitely empty areas are rejected. Fig. 5 shows the progression of the stored image from a totally filled area to the final image, as each angle updates the image a little more. Fig. 5(a) is the input image, the object that the sensor is looking at. Fig. 5(b) is the starting image. After obtaining data at the angle of 0°, we know that the object is confined to the narrow rectangle in Fig. 5(c). After taking data at the rotation of 90°, we can narrow the box further, as shown in Fig. 5(d). As the data-gathering continues, we can continue to update our image (Fig. 5(e-g)), until we find the final image, a convex cross-section of the object (Fig. 5(h)). In this example the input image is simulated, so the quantisation of the final image is partly due to the quantisation of the input image.

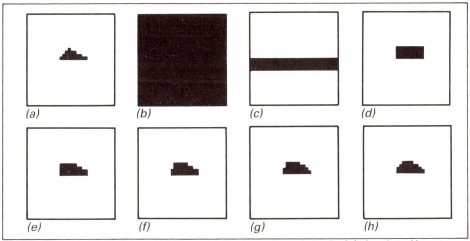

Fig. 5   *Progression of a stored image as each angle updates it: (a) simulated input image; (b) starting image, a totally filled area; (c) image after data taken at 0°; (d) image after data taken at 0°, 90°; (e) image after data taken at 0°, 90°, 30°; (f) image after data taken at 0°, 90°, 30°, 120°; (g) image after data taken at 0°, 90°, 30°, 120°, 60°; (h) the final image, after all six angles*

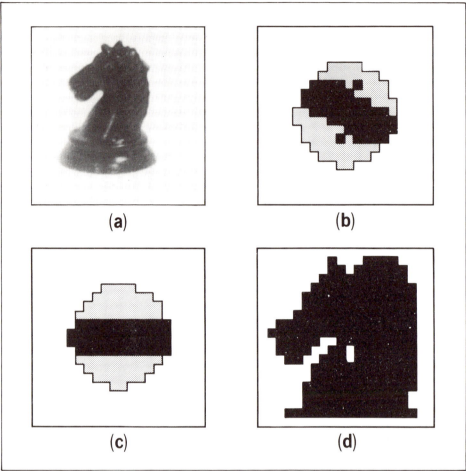

(a)

(b)

(c)

(d)

Fig. 6   Example illustrating sensor's location and orientation abilities: (a) real photograph of knight chess-piece, from a side view; (b,c) the two cross-sections are of the head and base; the darker one is higher; (b) cross-section of knight at a clockwise rotation of 30°; (c) cross-section of knight with no rotation; (d) linear scan of knight

In general, this method of image reconstruction leads to a polygon as the final image. This polygon is only an approximation to the real image, and a better approximation can be obtained as follows. By construction, we know that each side of the polygon must contain at least one point of the perimeter of the real image. It is easy to show that each polygon side contains a segment of the real-image perimeter whose real length, on the average, is one-third of the polygon side. By symmetry, this segment is centred on the polygon side. If we join all these segments we obtain a new polygon which is the best approximation to the real image.

An example of how the sensor can be used for combined position, orientation and shape detection is given in Fig. 6. Fig. 6(a) shows a photograph of a knight chess-piece that is approximately 1cm in height. After the robot has moved its hand to the general area of the knight, the sensor can determine two cross-sections (the head and the base) at different heights Z. Thus it determines the position and the orientation of the head (Fig. 6(b)). After establishing the orientation through standard pattern analysis algorithms, it is easy to align the fingers with respect to the head (Fig. 6(c)), and scan linearly along the Z-axis to form a two-dimensional silhouette (Fig. 6(d)). Thus, the three-dimensional shape of the knight can be reconstructed.

The sensor's accuracy is limited only by the width of the light beams that it uses and by the robot's accuracy. In the current implementation, the robot's accuracy is greater than the beam width, but this is not necessarily true for high-resolution beams (less than 50μm). Even if the light beams are separated by a distance larger than their width, the sensor can still cover the area between the light beams by having the robot shift its fingers a distance equal to the beam size. (Hence the requirement that the light beams be separated by a multiple of the beam width.) This shifting of the sensor, and thus the greater resolution of the sensor, is possible because of its dynamic nature.

Concluding remarks

We have designed and built an optical sensor that is operated in the dynamic sensing mode. It is a relatively simple construction that can be made modular and is easily installed on a wide variety of robots. It can locate objects to a precision of $\sim$0.5mm. This precision can be increased by building an integrated-optic array.

The sensor is not designed for competition with a video camera vision system. It does, however, have complementary features, e.g., it can work in the dark, it ignores inadequate lighting and background contrast, shadows and reflections, and it is mobile so it can access widely scattered or hidden parts.

A unique property of the sensor and a definite advantage over vision is that it can calibrate itself. All that is necessary for the calibration of each beam is a measurement of its intensities when it is uninterrupted and when it is totally interrupted. Measuring the unblocked beams is simple, and the robot can easily manœuvre its hand so that the beams are completely blocked by a wide calibration square. This self-calibration could also be used as a self-test. If the uninterrupted beams are not of a certain minimum intensity, then the sensor is not properly functional; this means that the light sources and their alignment with the corresponding photodetectors should be checked.

Perhaps the most important difference between dynamic sensing and vision is that the data acquisition strategy must be considered in the dynamic sensing algorithm. For example, the robot can find the position of the object by finding the smallest box containing it. This takes only two angles – $0°$ and $90°$ – so if it can accomplish the task with just that information, then it should not take further data.

These data acquisition strategy problems are not restricted to this sensor, and they will appear more often as robots acquire more sophisticated sensors. The general problem can be called 'intelligent machine perception', and it should be considered distinct from the problem of pattern analysis.

While emphasis on pattern recognition is appropriate for computer vision, we feel that for robots the problems of data acquisition strategy deserve further attention.

References

[1] Unser, M. and de Corlon, F. 1982. Detection of defects by texture monitoring in automatic visual inspection. In, *Proc. 2nd Int. Conf. on Robot Vision and Sensory Controls*, pp.27-38. IFS (Publications) Ltd, Bedford, UK.
[2] Beni, G., Hornak, L.A. and Hackwood, S. 1983. Proximity sensing using re-entrant loop magnetic effect. *Sensor Review*, 3(2): 68.
[3] Harmon, L.D 1983. Automated tactile sensing. *Int. J. Robotics*, 1: 3.
[4] Beni, G., Hackwood, S., Hornak, L.A. and Jackel, J.L. 1983. Dynamic sensing for robots: An analysis and implementation. *Int. J. Robotics*, 2: 51.
[5] Shneier, M., Nagalia, S., Albus, J. and Haar, R. 1982. *Workshop on Industrial Applications of Machine Vision*, Washington, DC, USA.
[6] Salisbury, J.K. 1982. Ph.D. Thesis, Stanford University, Stanford, CA, USA.

A PRACTICAL VISION SYSTEM FOR USE WITH BOWL FEEDERS

A.J. Cronshaw,
PA Technology, UK

W. B. Heginbotham
URB and Associates

and
A. Pugh
University of Hull, UK

A prototype system for the automatic visual recognition of small industrial components fed from a bowl feeder is described. The microprocessor at the heart of the vision system is capable of recognising different orientations of components and can also detect faulty components. The system is restricted to detecting tolerances of 5% or worse. The system would be suitable for feeding and sorting small components prior to automatic assembly. It can handle up to 40 components per minute. A novel feature of the system is the use of fibre optics in the imaging subsystem. This has enhanced the imaging capabilities to include both a side view and a plan view of the component under inspection.

The equipment to be described is a laboratory prototype system developed at the University of Nottingham, Department of Production Engineering where the authors were formerly engaged in research. Previous vision projects at Nottingham go back to the early 1970s with the SIRCH visually interactive robot[1], and more recently work by Pugh and Waddon[2].

The advantages of programmable visual inspection systems are increasingly being recognised. The paper will describe a vision system for the inspection of components as fed from a bowl feeders. This type of system would be attractive in short batch environments where the system can be reprogrammed quickly. It also offers the advantage over conventionally tooled bowl feeders in that it is possible to detect faulty components. Hence these can be rejected from the stream and the problem of machine jams is therefore greatly reduced.

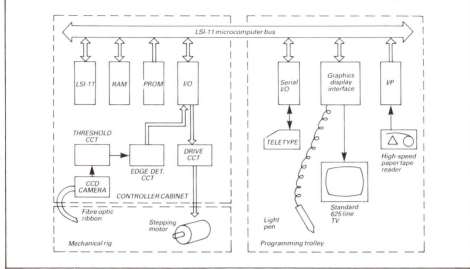

Fig. 1   Block diagram of the vision system

Overview of the vision system

The main parts of the vision system are shown in Fig. 1. The parts feeding system consists of a bowl feeder and an escapement to separate individual components as they emerge from the bowl. Components up to 2in. long by 0.75in. wide or high can be handled. Each individual component is then pushed forward through the viewing station. It is here that the visual image of the compnent is captured by the imaging system. Images of both the component's plan view and the component's side view are captured and passed to the microcomputer system for analysis. The component is then pushed forward ready for automatic handling by whatever device is to act on the flow of components. In the prototype equipment, the components were simply rejected into a bin ready for recirculation and the results of the microcomputer's analysis displayed on a VDU screen.

It is envisaged that in a real application, the flow of components would be selectively channelled into accept or reject bins, or passed into a device to actively orientate the component ready for automatic assembly. Alternatively, a robot could be used to pick up the parts. The result of the microcomputer's analysis would be passed to the robot to inform it of the nature of the component it is about to handle. Appropriate action could then be taken by the robot when gripping the component and also in using the component in the correct orientation.

An important facility of the vision system is the programming console through which the user is able to set up appropriate inspection procedures for a particular type of component under examination. The console consists of a VDU screen which is capable of graphical displays, and a light pen which the operator can use to point at menu options displayed on the screen. The normal procedure is to program the vision system (prior to automatic running), instructing the system with the aid of specimen good components which are shown to the system.

Fig. 2   Diagram of the imaging system

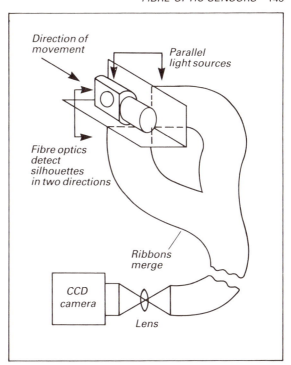

Direction of movement

Parallel light sources

Fibre optics detect silhouettes in two directions

Ribbons merge

CCD camera

Lens

Imaging system

The imaging system is shown in Fig. 2. The operation of the imaging system is as follows. Light is arranged to shine down vertically on the component in the viewing station so that a shadow is cast on the track floor. Fibre optics are set into the track floor in a line across the track. The fibres pick up a slit section of the component silhouette and transduce the image to a linear camera mounted a couple of feet away. Thus at any instant, the camera is able to see a thin slice of the component's silhouette. However, as the component is pushed forward through the viewing station, a series of such slices or scans is built up to give a two-dimensional plan view of the component.

Simultaneously, a side-view silhouette is built up in a similar fashion. To achieve this, light is also arranged to shine across the track so that a shadow of the side view of the component falls on the track side. Fibre optics are similarly arranged to capture the side-view silhouette.

An ingenious feature of the system is that the two ribbons of fibre optics share a single linear camera. (A 256×1 charge-coupled device camera was used.) The 256 elements of the linear camera can be allocated in any proportion to the side and plan view. So for example, the two views could each use 128 elements. However, if the component under consideration had relatively little height compared to its width, then it is possible to allocate say 56 elements to the side view and 200 to the plan view. This is illustrated by the graphical displays obtained for two different components in Fig. 3 and Fig. 6. The upper portion of the screen depicts the side view, and the lower portion depicts the plan view.

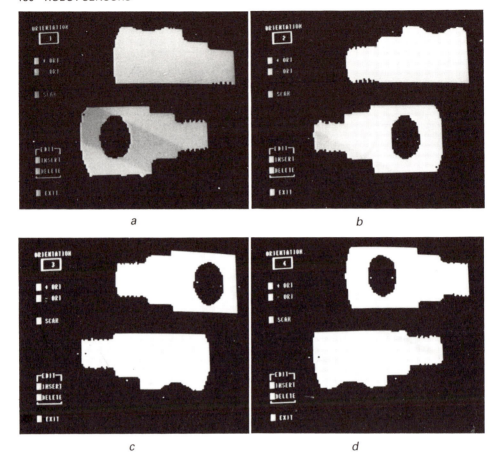

Fig. 3   *Four natural resting orientations of bicycle brake adjuster (courtesy of Raleigh Industries)*

It will be observed in these graphical displays that the side and plan views are offset slightly. This is a deliberate result of the design of the viewing station. The lighting/fibre optics for the side and plan views are not in the same plane. They are displaced approximately 0.25in. apart. This prevents lighting from one projection being reflected off shiny component surfaces into the other projection.

Using fibre optics in this way is a novel feature of the vision system. The combination of solid-state imaging and fibre optics has not been widely reported. However, a different application involving these two ingredients is reported by Coleman and Swinth[3].

The combined image is composed of a grid of 256 × 224 picture elements. The former is the number of camera elements, the latter is the number of scans captured. (This number of scans was found convenient to display on the graphics display unit.)

The camera signal is processed by various electronic circuits before being passed to the microcomputer. The grey-level signal is thresholded to give a binary-valued signal and hysteresis is incorporated to eliminate edge jitter.

A second circuit detects the position of black/white or white/black transitions and encodes the image data in terms of these edge positions. For simple image shapes, this results in a significant reduction in data volume to be input to the microcomputer.

Programming the system

In the set-up phase, the operator must instruct the system by showing it specimen good components, and by specifying detailed inspection procedures. The computer system stores the entered information as the master data. When later switched to the run mode, the computer will test component images against the master data in order to attempt component matching.

The computer's master data is organised as a series of templates corresponding to each discrete orientation of the component under consideration. It is assumed that the action of the bowl feeder will be to encourage components to take up discrete natural resting states. Only a small number of templates is therefore necessary to cover these well-defined component orientations.

For example, the brake adjuster component has four stable resting states which produce the graphical displays shown in Fig. 3. (This component has a flat machined along its length which tends to align itself against the track floor or against the track side.)

The system is programmed by interacting with it through the medium of a graphics display unit and a light pen facility. The system is able to display information and instructions to the operator on the TV screen, and the light pen can be used to enter commands to the system simply by pointing the pen at the relevant options displayed. This is believed to be a particularly powerful feature of the vision system since it allows very rapid specification of inspection procedures of a detailed nature. It is also a widely acceptable input/output device in these days of video games and teletext receivers. (Indeed Loughlin and Pugh[4] have used teletext equipment as part of a vision system.)

Typically, the operator would program the system for the brake adjuster as follows. A brake adjuster in the first orientation would be manually positioned in the escapement. The operator then uses the light pen to select menu option SCAN. The system responds by pushing the component through the viewing station in order to scan its image. The image is then displayed on the graphics screen (Fig. 3a). The operator must now position inspection points on the displayed image. He does this by penning INSERT (menu option) then by pointing the pen at some relevant point on the displayed profile.

For example, to check very simply for the existence of a hole in this particular component, a point could be inserted in the middle of the hole. The computer system stores the (x, y) coordinate of each inspection point along with its 'should-be' value which can be either 'inside' or 'outside' depending on whether the point is in the dark or light part of the silhouette. For practical purposes, as many points as desired can be inserted. (This is only limited by the size of the computer memory.)

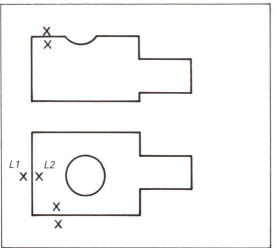

Fig. 4   Inspection points for length, height and width measurement

It is possible to use inspection points to check for presence of major characteristic features of a particular component orientation. It is also possible to use inspection points to check simple measurements. For example, the length of a component can be tested by inserting two inspection points L1 and L2, one just inside the profile and one just outside as shown in Fig. 4. Now the right-hand edge of the component will always be a constant reference (it rests against the pusher) so that point L1 will fail to be 'outside' if a component is too long, and point L2 will fail to be 'inside' if it is too short. Component widths and heights can similarly be checked (Fig. 4).

Features such as holes can also be checked for simple measurements. For example, a hole can be checked for minimum and maximum diameter and correct position by using three pairs of points (Fig. 5).

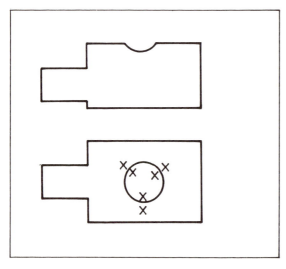

Fig. 5   Checking hole size and position

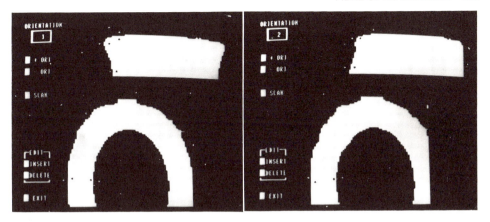

Fig. 6   Two orientations of brake seal

The tolerance of all inspection point measurements is determined by the operator since he can position points as tightly or as generously as he desires. However a limit of 5% has been found in practice, this being mainly due to the repeatability of the feeding system. The imaging system is otherwise able to resolve better than 1%. However, a 5% capability has proved to be adequate for many applications as will be illustrated below.

Applications

Case study 1 – Automotive brake seals

The rubber washers used in this experimental run are of the type used in automotive brake cylinders. The seals have specially shaped peripheries designed to slide against the inside cylinder wall. A batch of seals consisting of a mixture of various sizes and types was loaded into the bowl feeder to test the vision system's recognition capabilities.

A particular type of seal was selected to be programmed for recognition. It was of average size for the mixture, and actually quite difficult to differentiate with the naked eye from other similar types.

The imaging system was set up to suit the component's relatively small height compared to its width. The bevelled outside edge of this model of seal is clearly visible in the two graphical displays for the two orientations (Fig. 6).

Table 1   Results of brake seal test run

System decision	Observed orientation							
	Orientation 1	Orientation 2	Foreign type of seal	Two seals piggy back	Two seals side/side	Blank scans	Totals	%
Orientation 1	146	-	-	-	-	-	146	33
Orientation 2	-	86	-	-	-	-	86	20
Orientation 3	-	-	164	36	5	4	209	47
						Total	441	100%

Table 2   Brake adjuster (orientation) test run

System decision					Observed orientation					
Orientation	1	2	3	4	Other miscellaneous orientations	Two components	Grossly defective	No component	Total	%
Orientation 1	202	-	-	-	-	-	-	-	202	36
Orientation 2	-	272	-	-	-	-	-	-	272	49
Orientation 3	-	-	10	-	-	-	-	-	10	2
Orientation 4	-	-	-	14	-	-	-	-	14	2
Orientation 5	-	-	-	-	12	2	5	42	61	11
								Total	559	100%

Note that the plan view does not need to fully cover the entire ring – to measure the inside and outside diameters, only a generous semicircle is needed.

The inspection points to recognise the two orientations are shown in Fig. 6. A pair of points is used to check the inside diameter and another pair to check the outside diameter in the plan view. In the side view, the shape of the bevelling is checked by four inspection points and a further two points check the thickness of the seal (i.e. height). The inspection points are repeated for both of the orientations as required.

The system was then set into automatic operation and the performance of the system observed. The results are shown in Table 1. The system's decision could be one of three possibilities: orientation 1 or orientation 2 as already programmed, or orientation 3 if neither of the first two matched. The results show that for this test run of 440 scans, the system correctly recognised all orientation 1 and orientation 2 seals without error. More importantly, other types of seal were never labelled with these numbers. Instead they were labelled 'orientation', along with a few other error conditions such as two seals feeding together.

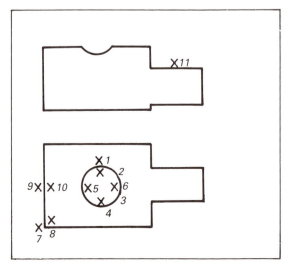

Fig. 7   Inspection points for brake adjuster (orientation 1)

Table 3 Results of brake adjuster inspection run

System decision	Observed orientation						
	Good component orientation 1	Defective components Gross Moderate Slight				Total	%
Orientation 1	457	-	-	not		457	91
Orientation 2	12	16	15	reliable		43	9
					Total	502	100%

Case study 2 – Bicycle brake adjuster

Two test runs were made using the bicycle brake adjuster component. The first run was made with the system programmed to recognise orientation only, without particular regard to detecting defective parts. The results shown in Table 2 confirm the system's integrity in this operation.

A more demanding exercise was then carried out to perform inspection of brake adjuster orientation 1 alone. The components were manually pre-orientated in orientation 1 so that the measurement capabilities of the system alone could be observed and recorded. Known likely defects in this component include incorrect overall length, excessive rounding of corners, and incorrect hole size/position. A batch of components with a wide spread of size defects was fortunately available. Accordingly, inspection points 1 were positioned as shown in Fig. 7. Experimental results are shown in Table 3. The system correctly asserted orientation 1 only when a good component was scanned. It did however reject a small number of good components (classifying them as orientation 2 which is used for error conditions) as well as rejecting all grossly defective and moderately defective components. 'Moderately defective' refers to components outside a 5% tolerance. Gross defects include scrap items or malformed components. 'Slightly defective' components are those within a 5% tolerance, but outside the spread of the majority of components in the batch which typically deviated by only 2.5% from a mean. The system was unable to reliably detect slightly defective components. The system was however reliably able to detect defective components with worse than 5% defects.

Software organisation

The software to support the vision system has been organised as a real-time multitasking system. This enables the microcomputer to simultaneously control the mechanical rig (including bowl feeder, escapement and pusher) and at the same time capture image data as well as perform image analysis.

The real-time language used to express this software is itself the author's own design[5]. Modular design of software routines has been combined with a medium-level language approach. The result is that the software is well organised and highly efficient.

Concluding remarks

The system has demonstrated itself to be effective for the visual recognition of component orientation and the detection of defective components.

Through the medium of the graphics display unit, a degree of human involvement in the set-up phase has been invoked which combines the skills of man and machine: man can devise strategies and specify relevant inspection procedures; the computer system will memorise these and repeat them indefinitely for a batch of components.

References

[1] Heginbotham, W.B. et al. 1973. The Nottingham SIRCH assembly robot. In, *Proc. 1st Conf. on Industrial Robot Technology.* IFS (Publications) Ltd, Bedford, UK.

[2] Pugh, A. and Waddon, K. The prospects for sensory arrays and microprocessing computers in manufacturing industry. *Radio & Electronic Eng.*, 47 (8/9).

[3] Coleman, W.J. and Swinth, K.L. 1978. Automatic high-speed process control/QA monitor. In, *Proc. 3rd Int. Conf. on Automatic Inspection and Product Control.* IFS (Publications) Ltd, Bedford, UK.

[4] Loughlin, C. and Pugh, A. 1979. Recognition and orientation of component profiles within the low resolution of teletext graphics. In, *Conf. Robots 79.* British Robot Association, Bedford, UK.

[5] Cronshaw, A.J., Heginbotham, W.B. and Pugh, A. 1979. Software techniques for an optically tooled bowl feeder. In, *Proc. Int. Conf. on Trends in On-line Computer Control Systems.* IEE, London.

4

Laser Sensors

Much work still needs to be done on the integration of laser technology with robotic workcells. The sensors are expensive and inevitably contain delicate mechanisms. The promise, however, is very good indeed.

A LASER-BASED SCANNING RANGE-FINDER FOR ROBOTIC APPLICATIONS

N. Nimrod, A. Margalith and H.W. Mergler
Case Western Reserve University, USA

A compact ($7 \times 7 \times 8$cm) scanning range-finder for robotic applications is presented in concept, implementation and performance. The system employs a new solid-state laser with internal beam conditioning. It measures range to a target in a plane along a measurement axis through a triangulation process. A transmitting mirror, rotating at a constant speed, serves to sweep a laser beam in a plane. A receiving mirror, rotating in synchronisation with the transmitting mirror, reflects the reference beam and the target-reflected beam through a focusing lens to a position-sensitive photodetector. The triangulation angle is extracted from the timing difference between the passage of the reference beam and target reflected beam across the centre point of the photodetector. Time to angle to range conversion is performed by a microprocessor. Resolution is a function of range, baseline length and variation in scanning speed. A typical precision might be 0.1% of range at 30mm and 0.07% of range at 1m.

During the past ten years considerable effort has been made in the development and design of industrial robots that can relieve humans from the burden of tedious or hazardous tasks, increase productivity and reduce production costs. A major issue in the development of robotic manipulators is the various kinds of sensors needed to obtain inputs from their environment. Techniques for tactile sensors, scene analysers, proximity sensors and range-finders, to name a few, have been investigated and tested. In particular, range-finders are needed to provide the controller of the manipulator with the sensory information required for functions such as object grasping, obstacle avoidance, moving parallel to a production table, and so on. A significant amount of work has been done at various locations in research and development of techniques for non-contact distance gauging. Optical, acoustic and magnetic sensors have been considered. Optical range-finders are relatively simple and can provide good accuracy in the range of interest, 1cm to 1m. The availability of solid-state continuous wave laser diodes, position-sensitive photodetectors and easily interfaceable microcomputers enables the design and implementation of compact and accurate optical range-finders for robotic applications.

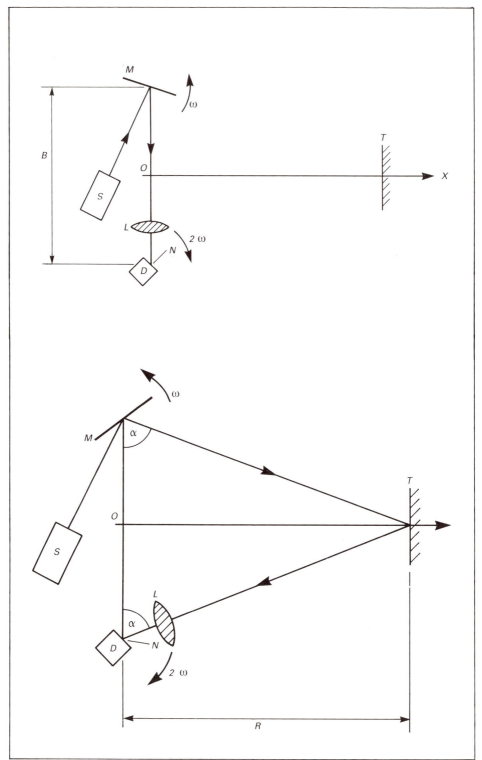

Fig. 1   Configuration of the basic triangulation laser-based scanning range-finder

Principle of operation

The basic configuration of the range-finder is depicted in Fig. 1. A laser source (S) generates a narrow collimated beam. The beam is aimed at a scanning mirror (M) such that the following conditions are fulfilled:

- The incident beam is normal to the axis of rotation of the mirror.
- The incident beam is directed through the axis of rotation of the mirror.
- The axis of rotation lies in the surface of the mirror.

The mirror rotates about its axis at a rate of ω in the direction shown in Fig. 1. Thus the reflected beam, according to the laws of reflection, scans the plane containing the incident beam at a rate of 2ω. This plane will be referred to as the 'measurement plane'. A position-sensitive photodetector (D) is placed in the measurement plane. This device is capable of generating a well-defined signal when a light spot crosses its null (centre) point (N). The distance between the axis of rotation of the mirror and the null point of the photodetector constitutes the baseline of the range-finder (B). A focusing lens (L) rotates around the photodetector such that its optical axis passes through the null point and it always lies in the measurement plane. The motion of the lens is synchronised with the motion of the mirror such that the angle α between the mirror-reflected beam and the baseline is always equal to the internal angle between the optical axis of the lens and the baseline (Fig. 1b). The synchronisation is achieved by proper angular alignment of both rotating elements and by having the lens revolve at a rate of -2ω, as shown in Fig. 1.

Consequently, the line along which the range is measured, referred to as the measurement axis (X), is the perpendicular bisector of the baseline. The range is measured in the half-plane which does not contain the laser source, and the intersection of the baseline with the measurement axis serves as the reference point (O) for the range measurement. Given that a target (T) of adequate reflectivity is located along the measurement axis within the range of acquisition, two null-crossing signals per mirror rotation cycle are generated by the photodetector. The first one occurs when the mirror reflects the incident beam directly along the baseline $\alpha = 0$), through the lens and onto the centre of the photodetector (Fig. 1a). The second one occurs when the mirror-reflected beam illuminates the target at its intersection with the measurement axis (Fig. 1b). At that moment, the optical axis of the lens also passes through the intersection point, and the lens forms the image of the illuminated spot on the null point of the photodetector. This is the 'triangulation instant'. Note that the target does not have to be specular for the distance measurement, as long as enough energy is incident on the photodetector.

The range R of the target is given by the trigonometric relation:

$$R = \frac{B}{2} \tan\alpha \tag{1}$$

where B is the baseline length. The triangulation angle α is directly

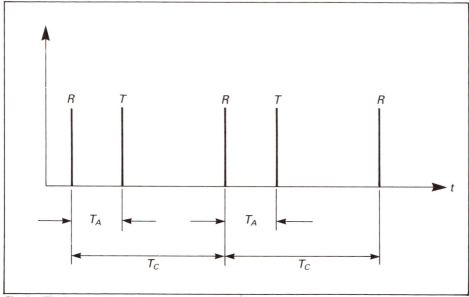

Fig. 2   Timing relationships between photodetector null-crossing pulses due to reference beams (R) and target-reflected beams (T)

proportional to the ratio of the time difference between the first and the second signals T_A, and the mirror scan cycle time T_C (see Fig. 2):

$$\alpha = 2\pi \frac{2T_A}{T_C} = 4\pi \frac{T_A}{T_C} \tag{2}$$

The extra factor of 2 in Eqn. (2) is due to the fact that as the mirror rotates by an angle α, the reflected beam sweeps an angle of 2α. Substituting Eqn. (2) into Eqn. (1) yields the range calculation formula for a single-surface mirror range-finder:

$$R = \frac{B}{2} \tan\left[4\pi \frac{T_A}{T_C}\right] \tag{3}$$

Assuming perfect mirror–lens synchronisation, it is of primary interest to investigate the effect of the length of the baseline on the sensitivity of the system. In order to resolve a small change $\triangle R$ in target range, the range-finder must resolve a small change $\triangle \alpha$ in the triangulation angle. These changes are sensed in the time domain as $\triangle T_A$. Solving Eqn. (3) for T_A gives:

$$T_A = \frac{T_C}{4\pi} \tan^{-1} \frac{2R}{B} \tag{4}$$

Differentiating with respect to the range R yields:

$$\frac{dT_A}{dR} = \frac{T_C}{2\pi B} \cdot \frac{1}{1 + (2R/B)^2} \tag{5}$$

Normalising by T_C gives:

$$\frac{1}{T_C} \frac{dT_A}{dR} = \frac{1}{2\pi B} \cdot \frac{1}{1 + (2R/B)^2} \tag{6}$$

Rewriting Eqn. (6) in a differential form gives the basic design equation:

$$\frac{\triangle T_A}{T_C} = \frac{1}{2\pi B} \cdot \frac{1}{1 + (2R/B)^2} \cdot \triangle R \tag{7}$$

Eqn. (7) provides, in terms of fraction of the cycle time, the time resolution that is needed in order to resolve a change $\triangle R$ at a range R, given the baseline length B. For example, a range-finder having a baseline of 5cm and a scan cycle time of 1s must resolve $1.98\mu s$ in order to achieve resolution of 1mm at 1m. This is within the realm of current digital timing techniques. The effect of the baseline on the sensitivity of the range-finder changes significantly as the range increases. For small ranges, where $R<<B$, the term $(2R/B)^2$ becomes much less than 1, and Eqn. (6) becomes:

$$\frac{1}{T_C} \frac{dT_A}{dR} = \frac{1}{2\pi B} \tag{8}$$

meaning that for small ranges the sensitivity is inversely proportional to B. However, for large ranges, where $R>>B$, the term $(2R/B)^2$ dominates the denominator of Eqn. (6), which becomes:

$$\frac{1}{T_C} \frac{dT_A}{dR} = \frac{B}{4\pi R^2} \tag{9}$$

indicating that the sensitivity is proportional to B. Nevertheless, the magnitude of the sensitivity is strongly diminished by the R^2 term of Eqn. (9). Table 1 summarises, numerically, a few cases and exemplifies the relationship discussed above.

An underlying assumption here is that the rotation rate is stable enough so as not to introduce a measurement error. Fortunately, only short-term stability is required, because the triangulation angle can be calculated from the ratio T_A/T_C and not just from the absolute value of T_A. Hence, the effect of long-term variation in the rotation rate over a few cycles will be cancelled in the ratio process. A synchronous ac motor with adequate inertia can ensure this required short-term stability.

Table 1   The time resolution (in μs) required to resolve 1mm at different ranges and baselines for scan cycle time of 1s

Range (cm)	Baseline (cm)				
	5	8	10	12	15
1	2744	1872	1530	1290	1042
10	187	274	318	351	382
100	1.98	3.18	3.97	4.76	5.93

The modified triangulation configuration

Motivated by the overall goal of compactness of the range-finder, the configuration shown in Fig. 3 has been developed. The principle of operation remains the same, but the receiving side of the system is modified. The rotating lens is replaced by a rotating mirror MR_2 which is at a 45° inclination with respect to the measurement plane.

Furthermore, the surface of MR_2, its axis of rotation and the measurement plane, all coincide at one stationary point (W), which corresponds to the null point of the photodetector in the basic configuration. Consequently, reflected beams which lie in the measurement plane intersect that point during triangulation instants. They are reflected by 90° and travel upward

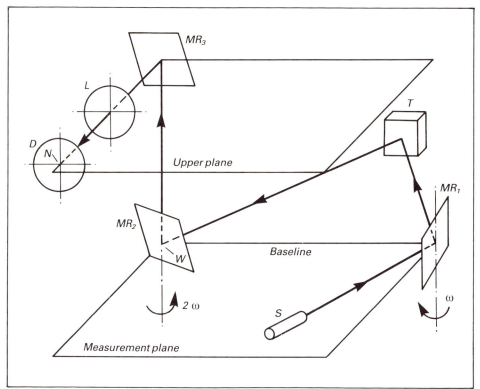

Fig. 3   Configuration of the modified triangulation range-finder

along the rotation axis of MR_2. A plane parallel to the measurement plane, referred to as the 'upper plane', is located above the measurement plane. The surface of a fixed mirror MR_3 is orientated such that it is parallel to the baseline, and it generates an angle of 45° with the measurement plane. As a result, beams reflected along the rotation axis of MR_2 are reflected again by 90° at MR_3 and travel in a path which lies in the upper plane, is perpendicular to the baseline and is directed opposite to the measurement axis. This path is referred to as the 'upper optical axis'. A fixed lens L is centred upon that optical axis, and the photodetector D is placed behind the lens, with the optical axis being normal to the detector's surface at its null point (N).

Implementation of the compact range-finder

A block diagram of the major parts of the system as implemented by this research is shown in Fig. 4. The range-finder assembly consists of a mechanical enclosure containing a laser diode, a photodetector, mirrors, a lens, and a motor with its associated gears. The laser driver circuit modulates the laser diode's radiation, and the phase detector circuit detects the null crossing by monitoring the photodetector output. The null crossing pulse (NCP) generator provides digital pulses which correspond to the null crossing instants, and these pulses are clocked by the timer board residing in the SBC 80/20 microcomputer. The CPU board, in conjunction with the maths board, executes the program which extracts the target range from the NCPs. Finally, the measured ranges are displayed in real-time on the CRT screen.

The complete range-finder design is broken into five separate yet coupled designs:

- The electro-optical part of the system, which consists of a laser source, transmitting and receiving mirrors, a focusing lens and a position-sensitive photodetector.

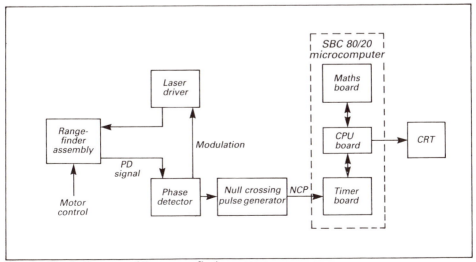

Fig. 4   Block diagram of the range-finder system

- The mechanical part of the range-finder, which includes the mechanical structure containing the electro-optical elements, the motor driving the rotating mirrors and the mechanics linking them.
- The computing hardware part of the system, which includes the digital timer circuit clocking the arrival of the NCPs, the microcomputer with its supporting maths board that performs the range calculation, and the CRT which displays the measured range.
- The analogue part of the system, which includes the laser driving circuitry, the circuitry that processes the signal generated by the photodetector and produces the null crossing pulses fed to the computing part.
- The computing software part of the system, which includes the programs that calculate ranges of individual targets and continuously displays them on the CRT, and is interrupt driven by the NCPs to read T_A and T_C values.

An excellent and complete description of the electro-optical and mechanical parts is given elsewhere[1]. For a detailed explanation of the analogue part, the reader is referred to Sahajdak[2] and Nimrod[1].

Microcomputer hardware

The computing facility employed by the range-finder is Intel's System 80/20-4. Details on this stand-alone microcomputer system are given elsewhere[3]. It is based upon the SBC 80/20-4 single-board computer which includes an 8080A CPU, 4K of static RAM, up to 8K of ROM, 48 programmable I/O lines, an RS232C communication interface, programmable eight-level vectored priority interrupt structure and bus drivers for memory and I/O expansion. The system's card cage can house up to four expansion boards which are easily accessible via the SBC 80/20-4 multibus. The range-finder system requires only two expansion boards: a floating point maths unit board referred to below, and the timer board discussed below.

The I/O board used to carry out the floating point calculations is Intel's SBC 310 high-speed mathematics unit[4]. This unit is capable of performing floating point add, subtract, divide, square and square root; fixed point integer multiply, divide and extended divide; conversions between fixed and floating point representations; and test, compare and argument exchange. Floating point operations are accurate to nine significant digits and typically execute in $100–200\mu s$.

Timer board design

The design goals of the timer board are:

- To measure the elapsed time for the reference beam NCP to the target reflected beam NCP (T_A) and also to the next reference NCP (T_C) with sufficient resolution.
- To latch the time count and request the microprocessor to read it by issuing an interrupt.
- To identify reference NCPs as such, inform the processor of that fact and immediately reset the time count.

Fig. 5 The range-finder assembly: (a) front view and (b) half-front view.

The final design parameters of the range-finder shown in Fig.5a and 5b are baseline length of 53.975mm and scan cycle time of 250ms. Substituting these values in the design Eqn. (7) and solving for the time resolution $\triangle T_A$ given $R = 1$m, $R = 1$mm we get $\triangle T_A = 0.5365\mu$s. This means that the digital counter must resolve 0.5μs or better and have a count modulo of 250ms or more. The reference constant clock provided on the 80/20-4 system multibus runs at 9.216MHz, or a cycle time of 108.5ns. The minimum number of counter stages required to count up to 250ms at input rate of 108.5ns is $\log_2 250 \times 10^{-3}/108.5 \times 10^{-9} = 22$. This relatively large number of stages coupled with the high frequency clock necessitates the use of synchronous counter, preferably with a carry look-ahead feature.

The reference clock is taken from the multibus as CCLK/, and is prescaled by two using a JK FF. The resulting 217ns clock drives a 24-state, synchronous binary counter consisting of six 4-bit counter chips (74163). The counters are cascaded using the carry look-ahead feature but the prescaling mentioned earlier is required because the 108.5ns clock is too fast for worst-case carry propagation throughout the whole counter. Consequently, the least significant bit represents 217ns, while the most significant bit used represents 0.91s (217ns $\times 2^{22}$). Assuming a reference NCP has just arrived and the counter starts from zero, it is counting up and its outputs are synchronously and continuously being latched into three 8-bit parallel registers (74199). At some later time, but no more than $T_C/4 = 62.5$ms for a valid target (triangulation angle between 0 and $\pi/2$) a target NCP arrives. It is latched in a JK FF which in turn: inhibits further loading to the parallel register, thus freezing the last latched count, which is actually T_A; and generates an interrupt request to the CPU by asserting INT3/ on the multibus.

The outputs of the parallel registers are routed directly to the input ports on the CPU board. The CPU services the interrupt request by reading the count value from these three ports and then it issues an interrupt acknowledge (INTACK) via bit 0 of an output port. This resets the JK FF, the registers resume latching counts and the next target can be clocked. Towards the end of the scan cycle time, as bit 20 of the counter, which corresponds to-a time interval of 227.54ms, is set, a D FF is also set. The NCP which arrives next is presumably the reference NCP, and that fact is conveyed to the CPU by setting the MSB of the high count byte to '1'. This bit is called REF NCP FLAG, and the software uses it to distinguish between the values of T_A and T_C. Simultaneously, the counter is reset by a clear pulse which is generated by the D and the JK FF's, and the cycle starts over.

Software design

The software is divided into two parts: a real-time (interrupt driven) input process and a number crunching process.

The range calculation algorithm

The two major design considerations of the range calculation algorithm involve speed and accuracy: it must be fast enough for real-time processing and accurate enough so as not to introduce a calculation error comparable to the measurement error. As discussed before, the desired accuracy of the system must be at least $0.5\mu s/250ms = 2 \times 10^{-6}$, hence a calculation error of 10^{-7} or less is acceptable. Recalling the range calculation of Eqn. (3):

$$R = \frac{B}{2}\tan\ 4\pi\left[\frac{T_A}{T_C}\right] \qquad (9)$$

we have to select a suitable approximation for the tangent function. An excellent source for computer approximation is given elsewhere[5]. The accuracy of the maths unit is about nine significant digits. The algorithm designated 4242 approximates the tangent function in the domain $[0,\pi/4]$ with a maximum absolute error of $10^{-8.2} = 6.3 \times 10^{-9}$ on the order of the machine's accuracy. This approximation is given by:

$$\tan\left[\frac{\pi}{4}X\right] = X\ \frac{P_{01}\cdot X^2 + P_{00}}{X^4 + Q_{01}\cdot X^2 + Q_{00}} \qquad (10)$$

where X belongs to $[0,1]$ and the coefficients are:

$$P_{00} = 211.8493696 \qquad P_{01} = -12.5288887$$

$$Q_{00} = 269.7350131 \qquad Q_{01} = -71.4145309$$

Writing the polynomials of Eqn. (10) according to Horner's rule yields:

$$\tan\left[\frac{\pi}{4}X\right] = \frac{X\cdot(P_{01}\cdot X^2 + P_{00})}{(X^2 + Q_{01}\cdot X + Q_{00})} \tag{11}$$

which requires three multiplications, three additions, and one division. However, if the argument is between $\pi/4$ and $\pi/2$ meaning that X belongs to [1,2], then domain reduction must be done. X is replaced by $X/2$ before substituting it in Eqn. (11), and the tangent of the original argument is given by:

$$\tan(X) = \frac{2\tan}{1 - \tan^2} \tag{12}$$

which requires additional multiplication, subtraction, squaring and division. A rough estimate of the worst-case calculation time of the tangent, using maximum execution times[4], comes up to 1.23ms, which is about 0.5% of the cycle time. Obviously, the algorithm is fast enough for this range-finder implementation. Faster scan rates may necessitate the use of the currently available maths calculator chips that implement directly the tangent function.

The range calculation algorithm is summarised in Fig. 6. Note that the equality between the arguments of the tangent function of Eqns. (3) and (11), namely:

$$4\pi \frac{T_A}{T_C} = \frac{\pi X}{4} X \tag{13}$$

can be solved for X, yielding:

$$X = 16 \frac{T_A}{T_C} \tag{14}$$

thus enabling the algorithm to work only in terms of X.

Program operation

The general structure of the program can be divided into three parts: initialisation, target input process and target output process. A high-level flow chart of the program is given in Fig. 7. The initialisation process initialises the various I/O peripherals that interact with the program and a number of software variables. Following initialisation, the sign on message 'RANGEFINDER' is output to the CRT and then the program attempts to get in sync with the timer board. It does so by polling the interrupt controller until a timer interrupt is received with the REF NCP flag set. The interrupt system is then enabled and a 'READY' message is output to the CRT prompting the user to commence with the range measurements.

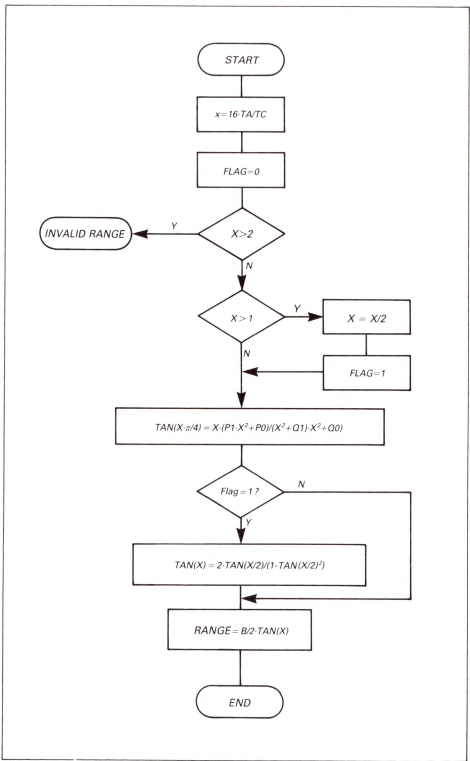

Fig. 6   Range calculation algorithm

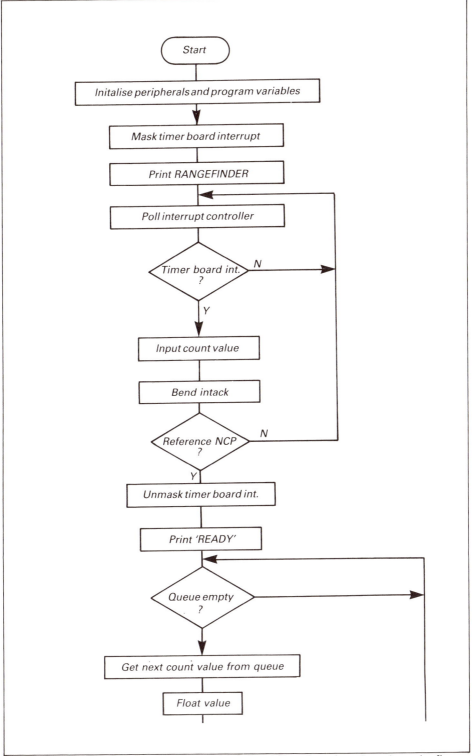

Fig. 7 Range calculation and display — high level flowchart (continued overleaf)

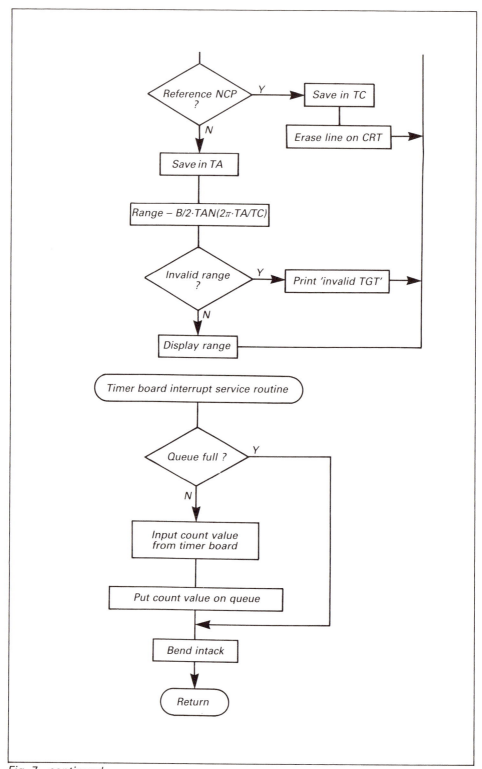

Fig. 7 continued

Since there is no connection between the rate at which targets are received and the rate at which they are processed and displayed, some form of buffering is required. The natural choice for the buffer is a circular FIFO queue[6]. For each NCP-induced interrupt, the associated interrupt service routine puts the 3 bytes of the count value from the timer board on the tail of the queue. The main program continuously checks to see if the queue is empty. If not, the next entry is removed from the head of the queue. If it corresponds to a reference NCP, the count value is used to update T_C for the next cycle. Otherwise the target range is calculated by the algorithm discussed earlier and displayed on the CRT in scientific rotation. Up to five targets can be displayed on one line. The queue management employed here discards incoming targets when the queue is full. No attempt is made to smooth the range values over successive cycles. However, with simple modification the program can be tailored to calculate and display the means and variances of the measurements for statistical purposes.

Concluding remarks

The basic principle developed in this research was shown to be feasible in an initial optical set up, which had the configuration depicted in Fig. 1. However, the prohibitively large size of that set up prompted the design and implementation of a compact optical head, based on the configuration of Fig. 3. Its dimensions were only $3.5 \times 3.5 \times 3.25$in. The performance of the analogue, digital and computational parts of the range-finder system met their design's goals.

Partial results were encouraging, as range readings of up to 160mm were obtained with standard deviations of 7.6mm. As the range shortens, the resolution improves dramatically. Based on our experiments we conclude that the measuring concept is sound and warrants continued refinement. Further work is required in the mirror rotating assemblies to minimise higher derivations in the parameter ω . Any variation in this quantity is interpreted as a $\triangle t$ and thus a change in measured range.

References

[1] Nimrod, N. 1981. A triangulation laser-based scanning range-finder for robotic applications. M.S. Thesis, Case Western Reserve University, Cleveland, OH, USA.
[2] Sahajdak, S. 1979. In-process optical gauging for numerical machine tool control and automated processes. Ph.D. Dissertation, Case Western Reserve University, Cleveland, OH, USA.
[3] Intel Corporation, 1978. *System 80/20-4 Microcomputer Hardware Reference Manual*. Intel Corporation, Santa Clara, CA, USA.
[4] Intel Corporation, 1977. *SBC 310 High Speed Mathematics Unit Hardware Reference Manual*. Intel Corporation, Santa Clara, CA, USA.
[5] Hart, J.F. 1968. *Computer Approximations*. John Wiley, New York.
[6] Harold, S.S. and Siewiorek, D.P. 1975. *Introduction to Computer Organization and Data Structures: PDP-11 Edition*. McGraw-Hill, New York.

LASER RANGE-FINDER BASED ON SYNCHRONISED SCANNERS

M. Rioux
National Research Council of Canada

A new geometrical arrangement is proposed to improve performance of optical triangulation.Two scanners in sychronisation allow a linear position to be used for surface topography measurement. Besides a large increase in speed of measurement, the new geometry allows considerable reduction of the optical head size compared with usual geometries, so the shadow effects are reduced proportionally. It also provides a means to obtain a very large field of view without compromising on resolution. Geometrical analysis and experimental results are presented.

For many years, much of the work in 3-D measurements involved passive techniques. One field of applications is earth topographic measurements where active techniques are impractical. Another one is scene analysis for military applications, where for obvious reasons of security, passive techniques are highly desirable. More recently the importance of 3-D vision in robotics was recognised, and research activities in this field are growing. The industrial environment provides new constraints and limitations to the applicability of usual techniques such as difficult environment, cost, and compactness, but, on the other hand, the proximity of objects allows active methods to be used to get 3-D data much more easily than passive techniques. Also, objects to be identified and manipulated are usually geometrically quite simple. Furthermore, objects of the same class are quite similar to each other (in contrast to natural objects in natural scenes) so that discrimination between classes should be easier than in natural scenes, especially with the use of 3-D measurements of their surface profile and some sort of compact geometrical description.

Active methods where a beam of light is superimposed to the naturally lightened scene greatly simplify the signal processing to be done to recover distance information. Among the various techniques described in the literature [1-15], we selected active triangulation as an attractive approach that has the potential to evolve toward a low-cost 3-D camera. Basic elements of such a system consist of a light source that is usually a laser, a scanning mechanism to project the light spot onto the object surface, and a position sensor with a collecting lens looking off-axis for the light spot. The distance measurement is done by trigonometric algebra applied to the projection direction (scanner angular position) and the detection direction determined by the light spot position on the sensor and the principal point of the collecting lens.

A large angular separation has to be used with usual triangulation technique to get a good compromise between resolution along range and field of view. The limitations imposed by a large angular separation are two-fold; first, continuous profile measurement is prevented by severe shadow effects, second, the optical head consisting of a scanner plus a camera is bulky. What we propose here is a new scanning mechanism that allows a compact camera head to be designed without compromising resolution and field of view and also limits shadow effects to a minimum. The basic idea is to synchronise the projection of a light spot with its detection. This is realised using two scanners instead of one. The first one projects a beam on the surface to be measured, and the second one is coupled to the receiving sensor to follow the spot.

Geometrical analysis

Usual triangulation geometry

Let us start the analysis with the usual triangulation geometry. In Fig. 1 we have a beam of light originating from position d along the X axis. The light beam projecting at an angle θ_0 defines a reference point $(d/2,l)$ that will be used for calibration purposes. At the origin there is a lens (not shown) of focal length f used to focus the light on a position sensor aligned parallel to the X axis and in focus at $-fl/(l-f)$ along the Z axis. We assume here that d, l and f are known and are, respectively, the distance between the scanner axis of rotation and the principal point of the lens, the distance between the

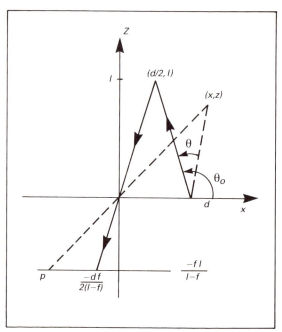

Fig. 1 Conventional triangulation geometry

common axis of projection and detection $(0,0 - d,0)$ and the reference point, and the focal length of the lens used to collect light from the scene. Under rotation of the scanner, the light beam rotates to another angular position $\theta_0 + \theta$ (θ is negative in Fig. 1). The spot of light on the position sensor moves from $df/2(l - f)$ to a new location p due to the intersection of the projected light beam with the object surface at x, z. By trigonometry we find the relation between the coordinates (x, z) and the parameters of the geometry as:

$$x = d \cdot p \left[p + \frac{f \cdot l \, (2l \cdot \tan \theta + d)}{(l - f) \, (d \cdot \tan \theta - 2l)} \right]^{-1} \tag{1}$$

$$z = -d \left[\frac{p(l - f)}{f \cdot l} + \frac{2l \cdot \tan \theta + d}{d \cdot \tan \theta - 2l} \right]^{-1} \tag{2}$$

Synchronised scanning geometry

Consider now a synchronised scanning geometry as shown in Fig. 2. The only difference here is the addition of a scanner which has its axis of rotation set at $(0,0)$. The scanner moves synchronously with the one at point $(d,0)$ (the projection scanner) and has the effect of cancelling its angular movement. The net result is to bring the position of the point on the sensor closer to the reference point $-df/2(l-f)$. That is illustrated in Fig. 2 as a trajectory from

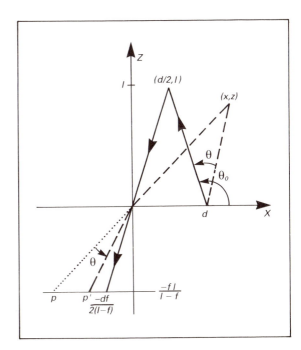

Fig. 2   Synchronised scanning geometry

point x, z to point p' and point p. Position p is the position that would be measured without synchronisation. Therefore, to find coordinates (x, z) with synchronisation, we relate p' to p, which is:

$$p = \left[p' + \frac{f \cdot l \tan\theta}{l - f} \right] \left[1 - \frac{p'(l-f) \cdot \tan\theta}{f \cdot l} \right]^{-1} \qquad (3)$$

and we use Eqns. (1) and (2). It can be seen from Fig. 2 that a change in the position of the light spot along the Z axis produces an equivalent angular shift for both geometries (usual and synchronised), but a change along the X axis produces a much smaller angular shift in the case of the synchronised geometry. In other terms, the position sensor is mainly used to measure range with synchronisation, while in usual geometries a large portion of the position sensor area is also used to measure x coordinates. That is interesting because the scanner mirror position can be precisely obtained. The net gain with synchronisation is that, with the same position sensor used for usual geometries, we can increase the focal length of the lens used to collect the light and obtain an increased resolution in range without reduction of the field of view. A more detailed description of the effect of synchronisation is described below.

Trajectory of the intersection of axes

In contrast to the usual triangulation geometry, a single point on the position sensor defines a surface in the object space. As an example, consider in reference to Fig. 2 that:

$$p = - \frac{df}{2(l - f)} \qquad (4)$$

for all angular rotation θ. This represents the trajectory of the intersection of the previously defined axis of detection (axis (1) in Fig. 3) and the projected light beam (2). As both scanners are synchronised, both axes rotate at the same angular speed (see Fig. 3). Axis (1) can be written as:

$$z^1 = x \tan(\theta - \theta_0) \qquad (5)$$

while axis (2) is:

$$z_2 = (x - d) \tan(\theta_0 + \theta) \qquad (6)$$

we know that,

$$\tan \theta_0 = -2l/d \qquad (7)$$

Fig. 3.   Trajectory of the intersection of axis

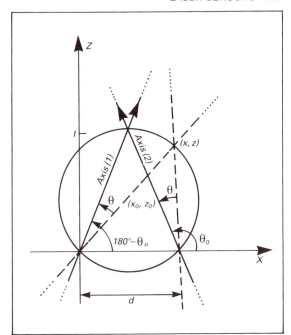

At the intersection of both axes,

$$z_1 = z_2 \tag{8}$$

This gives the trajectory of the intersection of the two axes as:

$$x^2 - dx - \left[\frac{l}{d} - \frac{d}{4l}\right] + dz + z^2 = 0 \tag{9}$$

This can be rewritten as:

$$(x - d/2)^2 + \left[z - \frac{l^2 - d^2/4}{2l}\right]^2 = \left[\frac{l^2 + d^2/4}{2l}\right]^2 \tag{10}$$

which is a circle of radius R;

$$R = \frac{l^2 + d^2/4}{2l} \tag{11}$$

centred at the coordinates:

$$x^0 = d/2 \tag{12}$$

$$z_0 = \frac{l^2 - d^2/4}{2l} \tag{13}$$

Note that the trajectory of the intersection of the axes is a circle inscribing the reference point and both scanners' axes of rotation.

For small angular separations between the projection and detection axes (which reduce shadow effects), the diameter of the circle is equal to ~l.

The fact that the trajectory of the intersection of axes is on a circle can be used advantageously for a class of inspection applications where profile measurements of spherically or near spherically shaped objects has to be made (rf antennas as an example). The advantage here is that low-resolution high-speed position sensors can be used to get very high-resolution profiles. This is because we know the reference surface, and we measure only the difference between the object's shape and the trajectory.

Bidmensional scanning

Consider now a dual-axis synchronised scanner as shown in Fig. 4. The geometry is similar to the one described in Fig. 2, but with the addition of a deflecting mirror which has its rotation axis parallel to the X axis and is located at a distance h along the Z axis from the synchronised line scanners. Angular rotation of this deflecting mirror is assumed to be known. Then the 3-D coordinates of a surface measured by the system are:

$$x' = d \cdot p \left[\frac{p + fl}{(l - f)\alpha} \right]^{-1} \tag{14}$$

$$y' = [h - \alpha(x' - d)] \sin \phi \tag{15}$$

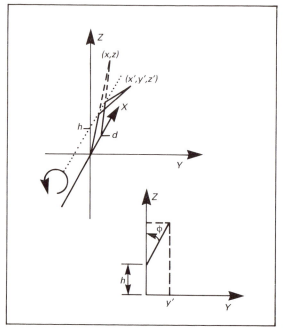

Fig. 4   *Dual-axis synchronised scanning geometry*

$$z' = h + [-h + \alpha x' - d]\cos\phi] \tag{16}$$

$$\alpha = \frac{d \cdot \tan\theta - 2l}{2l \cdot \tan\theta + d} \tag{17}$$

where ϕ is twice the rotation angle of the deflecting mirror. (Note that ϕ is negative in Fig. 4.) Then the reference surface for the trajectory of the intersection of the axes will be spherical if h equals z_0 [see Eqn. (13)]. For other values of h the reference surface is toroidally shaped with a radius of curvature given by Eqn. (11) along the X axis and a radius of $l - h$ along the Y axis.

Autosynchronised scanners

Scanner geometry

Two separate scanning mechanisms can be used to provide the rotation of projection and detection axes, but it requires critical adjustments to get stable synchronisation. A perfect synchronisation can be obtained using polygonal or pyramidal mirrors or a plane mirror coated on both sides. The idea is to use one facet of the scanner to project the light beam, and other facet of the same scanner for detection. Among the numerous geometrical arrangements, the one depicted in Fig. 5 has been implemented in our laboratory. It uses a pyramidal mirror consisting of six facets. A light beam from source S is deflected by a facet of the scanner to fixed mirror M_1. A

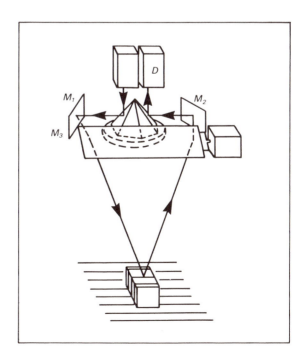

Fig. 5 Autosynchronised geometry using multifacet pyramidal mirror

second reflection occurs on deflecting mirror M_3. After scattering of the light onto the object's surface, the beam is deflected again on M_3 to another fixed Mirror M_2. Then we use the opposite facet of the pyramidal mirror to deflect the detection beam to position sensor D. Using a multifaceted mirror eliminates the need of trying to synchronise two scanners. The reference point discussed previously can easily be set to any desired value by angular adjustments of fixed mirror M_1 and M_2. Advantages of a pyramidal mirror for this application are (1) it allows a large collecting area for small inertia (compared to a polygon), and (2) angular coverage is a factor of two smaller than a polygon (with the same number of facets), which doubles the duty cycle. Deflecting mirror M_3 is scanning at a very low speed, typically two orders of magnitude slower than the pyramidal mirror.

Position sensor

Usual triangulation methods require a dual-axis position sensor to do surface profile measurements. One very interesting property of a synchronised approach is that it requires a linear position sensor only, which increases the bandwidth from ~10kHz to 1MHz. Also, the physical size of the position sensor can be much smaller due to subtraction of scan angles. Then to take advantage of the speed, we chose the smallest lateral effect photodiode available on the market. These are made of a photodiode deposited on top of a resistive layer. The electrical equivalent circuit of the device is shown in Fig. 6. The current produced by the photodiode is split by the resistive layer so that the position information can be simply obtained by dividing the difference of the currents by their sum. On the left-hand side of the figure, there is a physical description of the sensor along with the position signal characteristic as a light spot scans its surface. Interesting features of the lateral effect photodiode are insensitivity to large amounts of defocusing, a response time of 500ns, and an output position signal that gives the centroid of the incident light spot. Another property is that we can get a reflectance map of the scene under scan that is in the perfect registration

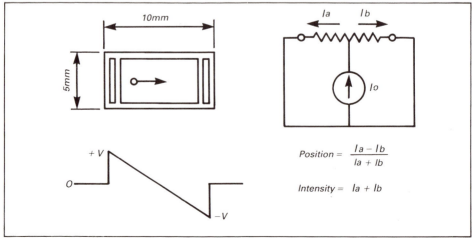

Fig. 6   *Lateral effect photodiode*

with the 3-D image. Indeed, the sum of the current is the amount of light diffused by the surface in the direction of the camera which gives a picture equivalent to what a TV camera provides. Interest of combined intensity and 3-D images have been discussed by Nitzan et al.[16].

Although CCD linear arrays are much slower (~10kHz), they provide higher resolution than lateral effect photodiodes at extra cost. The output signal from a CCD does not provide a direct measurement of the light spot position, and to take advantage of the high geometrical precision, it is necessary to use some kind of peak detection circuitry. They also provide high sensitivity allowing measurements to reach larger distances. As a comparison, the use of a lateral effect photodiode as a position sensor is limited to object–camera distances of the order of a metre using low-power lasers such as He–Ne and cw diode lasers. On the other hand, the use of CCD allows us to increase that distance to ~10m. Another comparison between the two sensors is to use the resolution-bandwidth product, which is, at a distance of 1m, of the order of 100MHz for lateral effect photodiodes (100 elements of resolution at 1MHz) and 40MHz for CCD using a peak detection algorithm having half a pixel of resolution.

A third type of position sensor is the dual-cells photodiode. Useless with usual triangulation geometries, this type of sensor can be used advantgeously with synchronised approaches. It is made of two small photodiodes located side by side. As for lateral effect photodiodes, the position signal is provided by the division of the difference of the currents generated by each cell by the sum of those currents. Its main advantage is a bandwidth of more than 30MHz (potential resolution-bandwidth product of ~1GHz). Signal processing circuitry required to cope with such a speed is the limiting factor of dual cells. It is also sensitive to geometrical distortion of the spot and has a very small displacement range. Nevertheless, its low cost makes it attractive for future development such as video rate 3-D cameras.

Beside these well-known position sensors, there are simple geometrical arrangements that were suggested in the literature[8,9] that provide high-speed high-resolution position sensing using a single photodetector. Such arrangements can also be used with synchronised geometries with the advantage of increased dynamic range along the Z axis.

Experimental results

An experimental phototype based on the schematic shown in Fig. 5 has been realised. A photograph of it is shown in Fig. 7. Although the design was made to use a diode laser, we got our first experimental results with a He–Ne laser of 15mW (not shown). At this time we are installing a RCA CDH-LOC diode laser. Exceptional beam quality and high cw power available from it (20mW) motivates our choice. The pyramidal scanner consists of six facets of 5cm^2 each. The angular speed adjustment ranges from 60 to 1800rad/s. This scanner provides X axis scanning, while a stepper motor driven flat mirror provides Y axis scanning. A collecting lens used for focus scattered light from the object's surface is shown on the schematic, fixed to the front end of the optical head. The lens has a 10cm focal length, and its output goes to a position sensor located on top in the rectangular black box. This box

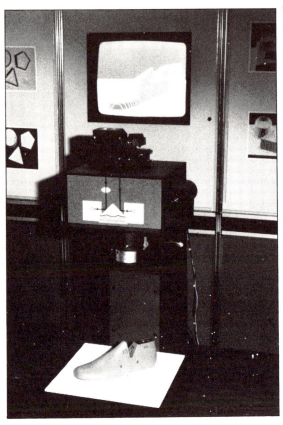

Fig. 7   Experimental prototype photograph showing the optical head, the object (shoe last) that has been scanned, and a display of the resulting measurements

contains the lateral effect photodiode plus two stages of amplification on both sensor outputs. Those signals are sampled at 600kHz, converted to digital using a 12-bit ADC, processed to obtain the position information, and sent to a memory.

We can see on the prototype table a shoe last that has been scanned. The 3-D data in memory are displayed on a TV monitor as an isometric view. Image size is 256 pixels/line, 128 lines. For clarity, we display only half of the lines on an isometric view. Scanning time is <1s, and numerical computation (difference divided by sum) is done in real-time. The pyramidal scanner rotation speed is set to 1000 rad/s. Our field of view is ~20°, so our spatial resolutions along the X, Y and Z axes are, respectively, 1, 2 and 0.4mm for a field of view of $250 \times 250 \times 100$mm.

A more detailed description of the camera output can be made using the photograph of Fig. 8. As mentioned earlier in the text, lateral effect photodiodes provide simultaneously an intensity image similar to what a TV camera would produce plus a tridimensional image. The first two pictures shown at the top of Fig. 8 are intensity images obtained by summing the two position sensor outputs for a shoe last and a human hand. The two pictures at centre show the 3-D signals as computed from the division of the difference by the sum of the position sensor outputs. Each pixel (voxel or surfel...) displays range information as grey levels, white being the shortest distance

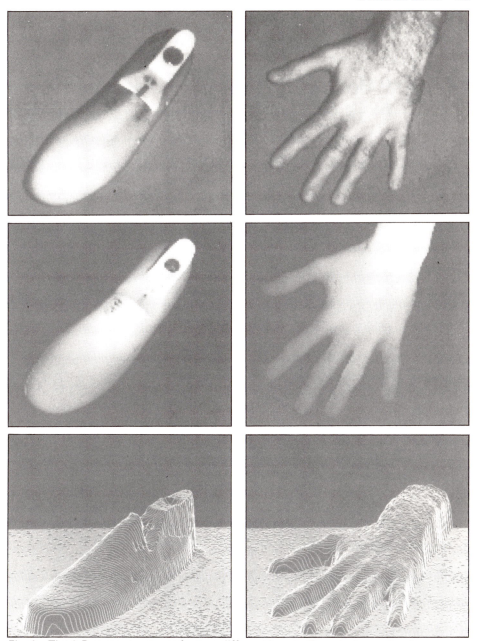

Fig. 8   The 3-D camera outputs for two different inputs. Top photographs show intensity output, while the four others show the 3-D data

and black the farthest. Our display is made on 32 levels only (5 bits), although the 3-D signal has a resolution of 8 bits. A more detailed view can be obtained using an isometric projection as shown at the lower part of the picture. The only inconvenience with this type of representation is that we have to hide a part of the available measurements to improve the quality of 3-D perception.

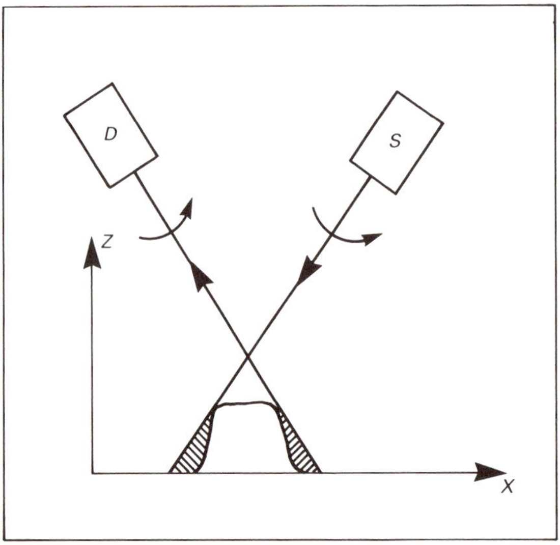

Fig. 9   Shadow effects

Discussion

The usual triangulation technique shows severe shadow effect problems associated with the requirement of a large separation between projection and detection axis in order to get a reasonable range resolution (Z axis). Shadow effects are of two types (see Fig. 9):

- The points on the surface that the projection beam cannot reach (points under the dashed zone on the left side in Fig. 9).
- The points on the surface that the position sensor cannot read (points under the dashed zone on the right side in Fig. 9).

Another way to get a higher resolution is to increase the focal length of the collecting lens, but then the field of view (along the X and Y axes) is proportionally reduced. Practically the usual triangulation geometry will use an angular separation of ~30°, which results in a very bulky optical arrangement. In contrast, synchronisation of scannings removes the tight relationship between X, Y and Z parameters. Scanner characterstics define the field of view, while the position sensor is mostly used to monitor range. In those conditions, we can bring the projection and the detection axis as close as we can without any reduction of the field of view. Then the resolution along Z can be set by the focal length of the collecting lens without any limitations on the other two variables. Our experimental prototype has an angular separation of 10°, which is limited by the size of the pyramidal mirror. Smaller angular separations can be obtained using polygonal scanners.

Looking back at Fig. 8, we see that the shadow effect represents a very small proportion of the image, although the surface under measurement shows a very broad range of surface orientations. In an attempt to reduce the shadow effect, we have to keep in mind that the removing of all shadows is a very difficult task. Points that are not illuminated by the projecting beam cannot be read in any geometry. The only way to get around this difficulty is

Fig. 10   Shadow effects reduction using two sensors

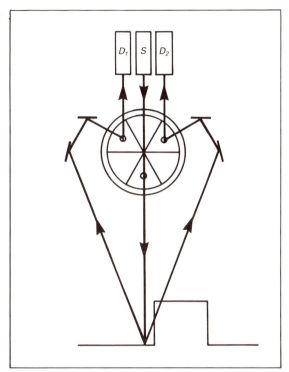

either to rotate the object or translate the camera head. If not practical, a more complex geometry as shown in Fig. 10 can be used to get measurements everywhere the projecting beam can reach (except for small deep holes). The arrangement is based on the use of two sensors looking on either side of the projecting beam while keeping synchronisation. Notice that there is a 90° rotation of the axis at the pyramidal mirror surface (top view) on Fig. 10. For rotating objects, a geometry is suggested in Fig. 11. Here the reference point is set to be on the rotation axis of the object. Because of the rotation of the object, there is no need for deflecting mirror, so the optical arrangement can be very simple and compact using polygonal scanner.

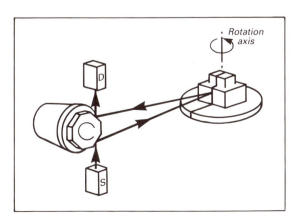

Fig. 11   Measurements on a rotating object

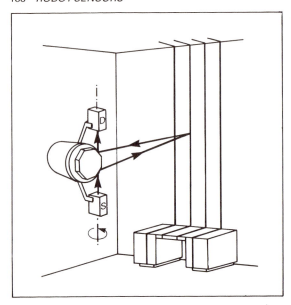

Fig. 12   A 360° range camera

The camera head can also be rotated as shown in Fig. 12 to map the interior of a room. Such an arrangement can be useful in robot guidance, where 360° scanning is attractive. One of the objectives of this project is related to robotic research, in particular, automatic assembly. Having the possibility to grab 3-D data of a scene in a fraction of a second has many practical advantages, especially in an industrial environment. Among them is a reduction in interference caused by mechanical vibrations. Furthermore, moving objects can be measured. Coupled to a robot arm, a 3-D camera can be used in conjunction with an appropriate vision processor to identify parts in the scene. On the other hand, we have to remember that a 3-D camera provides only the raw data (with noise), and there is still a lot of fundamental research to be done to use the 3-D data efficiently for object recognition. At least, one advantage of a 3-D camera is that the measurement of the object geometry is not a projection (as 2-D imagery). Indeed, we are actually measuring without ambiguity the surface profile of the objects under examination.

In terms of signal characteristics, the output from a 3-D camera as described here is very similar to a TV signal. In a 3-D camera the only difference is that the amplitude of the signal is related to geometrical charactersitics of the object. In contrast, 2-D cameras (like ordinary TV cameras) provide an output signal amplitude that is not geometrically related to the object; instead it represents the surface reflectance properties of the object combined with the ambient light conditions of orientation, intensity, and spectral characteristics. In addition, the net result is affected by the orientation of the object and by the proximity of other objects. This is why 3-D feature extraction from a 2-D image of a scene is so difficult to realise. A 3-D camera using an active approach also provides good immunity to background illumination. This is done by using an interference filter that allows only the laser light to reach the sensor.

Fig. 13 Use of an acousto-optic deflector to produce a reference surface of any shape

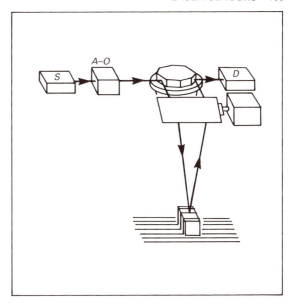

Resolution needed for object recognition is expected to be low (~1% in each axis). In contrast, inspection of manufactured objects, which is a big potential market for 3-D cameras, requires usually much higher resolutions. This can be met by CCD linear arrays, when high-precision measurements are needed, or by an arrangement such as shown on Fig. 13, when comparison of an object has to be made with a reference (e.g. inspection of pcbs for missing parts). This optical arrangement uses an acousto-optic deflector to change the angular orientation of the projecting beam prior to synchronisation, so we can program a reference surface of any shape for inspection. It means that, in normal operation where the inspected object is similar to the reference, the spot position on the sensor is stable. Now, if any part of the object is missing or distorted, a change in the position of the spot occurs. The departure of the new position from the stable one is proportional to the error and can be used to identify the missing part or the nature of the distortion.

Concluding remarks

A laser range-finder based on triangulation has been described. New geometries are suggested using synchronised scanners. Experimental results show a response time shorter than a microsecond using a lateral effect photodiode as a position sensor. This sensor also provides an intensity image which is in perfect registration with the 3-D one. Synchronised scanning allows very large fields of view to be obtained without compromising resolution along range axis. With the increasing development of diode lasers and holographic scanners, the new geometries proposed should lead to the realisation of compact low-cost high-speed 3-D cameras. We see besides robotic applications, inspection of manufactured parts, as well as medical application such as prostheses fabrication and diagnostic instrumentation.

Acknowledgements

The author wishes to thank P. Chartier and J. Domey for many discussions and help during the implementation of the 3-D camera system. Thanks are also due to L. Cournoyer and R. Misner for their excellent technical support in realising the programming and constructing the experimental set-up.

References

[1] Bodlaj, V. and Klement, E. 1976. Remote measurement of distance and thickness using a deflected laser. *Appl. Opt.* 15: 432.

[2] Boissonnat, J.D. and Germain, F. 1981.A new approach to the problem of acquiring randomly orientated workpieces of a bin. In, *Proc. 7th Int. Joint Conf. on Artificial Intelligence*, pp. 796-802.

[3] Altschuler, M. D. et al. 1981. Laser electro-optic system for rapid three-dimensional (3D) topographic mapping of surfaces. *Opt. Eng.*, 20: 953.

[4] Bernasconi, M. et al.1980. Accuracy of measurement through stereo images. In, *Proc. 5th Int. Conf. on Automated Inspection and Product Control*, p. 309. IFS (Publications) Ltd, Bedford, UK.

[5] Bien, F. et al. 1981. Absolute distance measurements by variable wavelength interferometry. *Appl. Opt.*, 20: 400.

[6] Burry, J. M. 1982. Contouring in real-time with Moire interference. *Opt. Commun.*, 41: 243.

[7] Fuzessy, Z. and Abramson, N. 1982. Measurements of 3-D displacement: Sandwich holography and regulated path length interferometry. *Appl. Opt.* 21: 260.

[8] Himmel, D. P. 1978. A laser measuring system for automatic industrial inspection. In, *Proc. 4th Int. Joint Conf. on Pattern Recognition*, pp. 952-954. IEEE New York.

[9] Indebetou, G. 1979. A simple optical non-contact profilometer. *Opt. Eng.* 18: 63.

[10] Morander, K. E. 1980. The optocator. A high precision, non-contacting system for dimension and surface measurement and control. In, *Proc. 5th Int. Conf. on Automated Inspection and Product Control*, pp. 393-396. IFS (Publications) Ltd, Bedford, UK.

[11] Page, C. J. and Hassan, H. 1981. Non-contact inspection of complex components using a rangefinder vision system. In, *Proc. 1st Int. Conf. on Robot Vision and Sensory Controls*, pp. 245-254. IFS (Publications) Ltd, Bedford, UK.

[12] Pipitone, F.1979. A ranging camera for 3-D object recognition. In, *Midwest Symp. on Circuits and Systems*, pp. 339-343.

[13] Sawatari, T. and Zipin, R. B. 1979. Optical profile transducer. *Opt. Eng.*, 18: 222.

[14] Taboada, J. 1981. Coherent optical methods for applications in robot visual sensing. *Proc. Soc. Photo-Opt. Instrum. Eng.* 283: 25.

[15] Kanade, T. and Asada, H. 1981. Non-contact visual three-dimensional ranging devices. *Proc. Soc. Photo-Opt. Instrum. Eng.* 283: 48.

[16] Nitzan, D. et al. 1977. The measurement and use of registered reflectance and range data in scene analysis. *Proc. IEEE*, 65: 206.

THE ORIENTATION OF DIFFICULT COMPONENTS FOR AUTOMATIC ASSEMBLY

C.J. Page and H. Hassan
Coventry Lanchester Polytechnic, UK

The aim of the work described in this paper is to develop techniques for the automatic orientation of complex three-dimensional components in parts feeders supplying automatic assembly machines. Computer-controlled inspection and handling of the parts is described and a range-finder imaging system for inspecting the components from all sides is developed. Software algorithms are also discussed.

The classes of part under consideration are those which are difficult or impossible to orientate with a sufficient degree of reliability using conventional, mechanical methods. A typical example is a component in the shape of a perfect cube and with a small tapped hole located off-centre in one of its faces. This part is difficult to orientate using mechanical methods, particularly if the hole is too small to catch on sensing pins or to make a measurable difference to the position of the component's centre of gravity. One traditional solution is the redesign of the part to include non-functional features specifically for orientation. Other methods are the magazining of difficult parts, or manufacture on the assembly machine itself. These techniques can be expensive, however, especially if production in medium-to-small batches is required.

The alternatives which are being developed are cheaper by virtue of their flexibility and ease of reprogramming to accommodate design or job changes. These methods use computer-controlled imaging and robotic techniques for recognition of component posture and subsequent reorientation.

The overall operation of the system is as follows. Parts are presented to a computer-controlled sensing head on a feed track from a conventional parts feeder. The component receptacle of the sensing head is a flat, transparent plate on which the part comes to rest in a random position and orientation and in any of its stable postures. The component is imaged simultaneously from each of the six sides required to perform a complete inspection of it from all directions. This requires a set of six identical sensors, each of which produces an image of the part from its own particular viewpoint. The data from the sensors is then analysed and correlated by the control microprocessor to yield a representation of the component's geometric shape.

This is compared with a preprogrammed version showing the part in its preferred orientation. Information from this comparison is then used by the microprocessor to derive information which will enable an industrial robot to pick up the part, reorientate it and place it on another feed track or on the assembly machine itself.

There is currently a great amount of research effort being devoted to computer-aided inspection, and particularly to robot vision and its application to automated assembly. Simple visual capabilities for industrial robots handling engineering components in manufacturing environments have been the subject of detailed study for the past decade or so[1,2]. More recently, major manufacturers have been showing serious interest in robot vision, and some are now developing their own application-orientated systems[3] (see also page 213). The results of robot vision research have also been applied to parts feeding equipment in work which is particularly relevant to the subject of this paper[4].

All these approaches employ a sensing device (usually a television camera of some form) which generates a two-dimensional image of the scene of objects before it. If depth information is required, ancillary sensors must be used. However, other workers have reported systems in which three-dimensional imaging is performed directly. Triangulation methods are most commonly used[5], although researchers at SRI have reported work on optical radar techniques applied to the imaging and analysis of office scenes[6,7]. Descriptions of research on the use of tactile methods for sensing three-dimensional engineering components have also been published[8].

Each of the sensors in this application generates a three-dimensional image of the workpiece from its own particular viewpoint by measuring the range to the part over a square matrix of sampling or picture points. Complete, unambiguous data on the workpiece is obtained, and this data is unaffected by variations in the angle and level of illumination. The method used for range-finding is similar to that used by the SRI workers[6,7], and measures range by time-of-flight measurements of a beam of laser light reflected back to its source by the component. This method is being employed in preference to triangulation because the latter can produce blind spots at the abrupt discontinuities in profile which occur in many engineering components.

The imaging system

The overall arrangement of the inspection station is shown in Fig. 1. Six identical sensors view the component simultaneously from each side. Each sensor scans a square field of view with a television-like master scanning system. The range to the object is sampled along each line of the scan to produce a square matrix of range values for each of the six viewpoints. An important feature is that the scanning beam is always normal to the image plane. This produces an image of constant size and one which is free from perspective distortion irrespective of the position of the component. The six sensors are arranged so that their fields of view intersect to produce an imaged volume in the form of a cube. The dimensions of the sensors are such that this cube has sides of length 100mm. The part under scrutiny rests on a

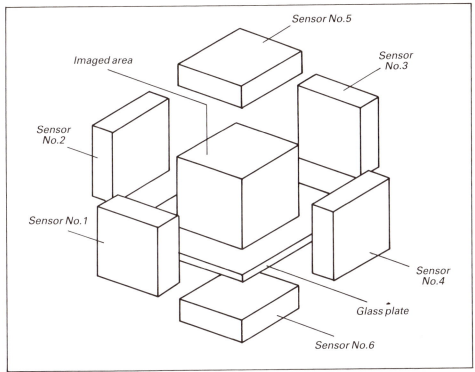

Fig. 1   Arrangement of inspection station

flat, anti-reflective coated optical glass plate. Reflection and refraction of the image 'seen' by sensor 6 through the transparent glass baseplate is minimised by the normal scanning arrangement.

The range to the component on the baseplate is sampled 64 times along each scanned line of the sensor raster.

To maintain symmetry, this requires that the raster scan consists of 64 lines. The distance between adjacent picture points is then 1.5625mm. Again to maintain symmetry, range must be measured in increments of 1.5625mm, giving 64 discrete range increments over the cubic 'image volume'. This means in turn that the image volume is divided up into 64^3 (262,144) elemental cubes of side 1.5625mm. If range is stored as a displacement from the front face of the cube with respect to each sensor, six bits are needed for each range value. If these are then stored one per byte in the memory of the supervisory microprocessor, a total of 4Kbytes of storage is required for each sensor, making a total of 24Kbytes in all. The control processor has a working memory of 56Kbytes, so there are 32Kbytes left for data processing software before time-sharing via backing storage media need be considered.

To be industrially viable, total scan time for the range sensors must be of the order of one or two seconds. This means that range must be sampled, calculated and stored about once every 500ms if a maximum imaging time of two seconds is fixed upon. This would seem to be achievable in a practical system while maintaining a reasonable compromise between accuracy, repeatability, durability, complexity and, of course, cost.

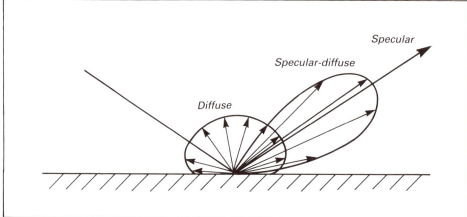

Fig. 2   Reflected light components

Range detection

The method used for measuring range to the object under scrutiny relies on detecting the light scattered from a small, intense spot of light projected onto the surface of the component. The light detector is positioned so that it measures scattered light in a direction coaxial with the transmitted beam, thereby eliminating blind spots which can be encountered when using triangulation methods. Because of the coaxial arrangement of transmitter and detector, the latter will be relying on picking up radiation from diffuse rather than specular, or mirror-like reflection at the surface.

For most engineering surfaces, reflection is a mixture of specular and diffuse reflection rather than just one or the other. This is shown in Fig. 2.

In practice, the amount of scattered light and the relative magnitudes of the specular, specular-diffuse and diffuse components are dependent on radiation wavelength, surface composition, and surface roughness. For instance, metals tend to be more reflective than non-metals, and this reflectivity increases with radiation wavelength. For engineering components, surface roughness is a particularly important factor in determining the amount of light scattered from the surface. The specular-diffuse and diffuse components increase from zero for a truly mirror-like surface up to significant levels for machined surfaces.

To illustrate these points, some simple experiments have been performed, in which a small spot of light is projected from a low-power helium–neon laser ($\lambda = 632.8$nm) onto various surfaces. Reflected radiation is detected by a silicon photodiode–amplifier combination which is situated as near coaxially with the incident beam as possible, and is measured over a range of angles of incidence of the laser beam for a number of engineering materials with different surface roughnesses. The general trends shown by these results are illustrated in Fig. 3.

For mild steel with good to average finishes (CLA values 0.08–1.5μm) produced by methods such as polishing, grinding, milling and turning, the response tends to be specular as expected (Fig. 3a), with a peak response varying from 70mV down to about 15mV. More importantly, however, for

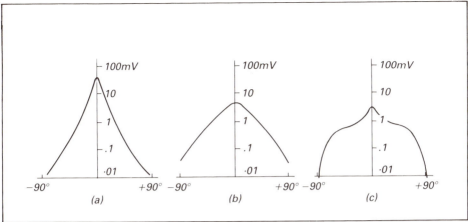

Fig. 3   General shapes of reflected light response curves for some engineering materials: (a) metals (good finishes), (b) metals (average finishes), and (c) non-metals (good finishes)

the purposes of the techniques being developed, a detectable output of at least 10μV is obtained at angles of incidence in excess of 80°. For rougher steel surfaces (CLA values > 2μm), specular-diffuse and diffuse reflection tends to dominate the response as shown in Fig. 3b. Here peak response is several millivolts, while the detector output at large angles of incidence (> 80°) is somewhat higher than in the previous case, being some tens of microvolts.

As mentioned above, non-metallic materials such as plastics, wood etc. are not so reflective as metals, and this is borne out by the results (illustrated in general form in Fig. 3c). The magnitude of the specular component is influenced by surface roughness and peak values range from 2 to 20mV. The response at large angles of incidence is similar to that for metals with average to rough finishes.

These experiments show that the basic imaging technique outlined earlier is viable for the majority of common engineering materials in that reflected radiation coaxial with the incident beam can be detected even when the surface is nearly parallel with the beam. Components with a truly mirror-like surface are obviously not a practical proposition as the amount of diffuse reflection is almost zero. Neither are parts made from translucent or transparent material, although for slightly different reasons. In these cases, an appreciable amount of the incident light will pass through the front surface of the component and be reflected from the rear face, increasing the optical path length and hence causing an error in the measured range value.

For this type of application, the choice of light source and detector is particularly important. Ideally, both should be small, robust, efficient, low-power devices. For the light source for each sensor a Ga Al As (gallium aluminium arsenide) solid-state double-heterostructure laser diode has been specified. This produces 15mW of optical output power with a 2V forward voltage drop and a typical operating current of 150mA. The output is more than adequate for this application even allowing for losses in the associated optical system. The wavelength of the optical output is 800nm, which is in the infrared. The laser diode chosen is a type specifically designed for continuous operation and for high-frequency modulation up to the gigahertz range.

The detector being employed for each sensor is a silicon avalanche photodiode. This is a solid-state device which incorporates internal gain (around 200 typically), and which can detect infrared light modulated with frequencies up to the gigahertz range. Its main disadvantage is the high dc bias voltage of around 200V which is required. Bias circuits can be designed to incorporate automatic gain control (agc) to protect the photodiode against electrical damage due to excessive optical power levels.

As mentioned earlier, range is measured by modulating the output from the solid-state laser with a sinusoidal waveform and detecting the phase shift over the path length of the light beam, which is twice the range. It is also possible to measure range by measuring the elapsed time between transmitted and reflected pulses of laser light. This is the method used for long range (several kilometres) work because the high power levels required necessitate the use of pulsed lasers operating at low duty factors. Both methods are possible in this short-range and low-power application. However, the continuous wave (CW) technique has several important advantages. Firstly, there is a turn-on, or lasing delay of one or two nanoseconds in the pulsed system. This can be reduced to zero for practical purposes by dc biasing the laser diode and superimposing a sinusoidal modulating waveform. Secondly, in the pulsed system the fast edges in the reflected light can be masked by the inevitable pick-up and noise. The effects of noise can be reduced in the CW system by passing the detector output

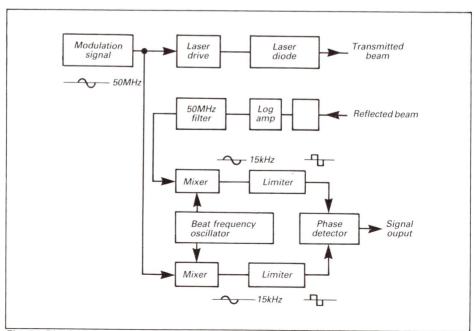

Fig. 4   Block diagram of range detection system

through a narrowband filter tuned to the modulation frequency. Thirdly, the detector output in the CW system can be mixed to generate an intermediate frequency (IF) signal with the same relative phase shift as the original, but at a much lower and much more tractable frequency.

The resolution required from the range detection system is approximately 1.5mm. At a modulating frequency of 50MHz, for example, this gives an incremental phase shift of 0.2°. While it is possible to measure much smaller values of phase shift, in view of the low signal levels and also the wide dynamic range of the signal it has been considered advisable to use a modulating frequency in the tens of megahertz range. The sensors are each 100mm away from the front face of the image volume, so this gives maximum and minimum signal path lengths of 400mm and 200mm, respectively. The corresponding phase shifts relative to a reference over these distances are 24° and 12°.

A block diagram of the electronic control system of the range-finder is shown in Fig. 4.

The 50MHz sinusoidal modulating signal is derived from a crystal controlled oscillator and modulates the laser diode bias and drive circuit directly. The laser diode transmits a beam of infrared light which is modulated at 50MHz with as high a modulation factor as possible to improve the signal-to-noise ratio at the detector. Reflected infrared light is picked up and amplified internally by the silicon avalanche photodiode detector. A logarithmic amplifier is used in the following amplification stage to compress the wide dynamic range of the incoming light levels. The practical results discussed earlier in this section show that this can be nearly 80dB. An additional factor in the practical range-finder system is the variation of detected radiation intensity with range. This obeys an inverse square law, which means that the total dynamic range of the signal could be as high as 90dB. The output from the amplifier is then filtered by a narrowband filter to reduce unwanted noise levels. A beat frequency oscillator is used to generate an intermediate frequency output of about 15kHz from both the initial, reference modulation signal and the output from the filter. After passing through a limiting stage, the phase difference between the two signals is detected and converted to digital form ready for interfacing to the control processor.

Obviously, there will be differential phase delays between the reference and received signals additional to those which are proportional to range. These are caused by phase delays in the detector chain prior to the mixer stage and also by the length of the optical path through the sensor. They are dealt with in a preprocessing operation by the processor, in which suitable offsets are subtracted from the output of the phase detector. Values of these offsets are stored in the form of a look-up table, and are derived from the results of a calibration operation in which the range-finder output is measured for surfaces at known ranges. Drift in the operating characteristics of the range-finder circuitry can also be compensated for by a variation of this method by incorporating at certain positions in the field of view reference points of known range which can be used to calculate a correction factor.

Scanning system

The beam scanning system uses a single transmitter and detector per sensor with a mechanical/optical arrangement performing the scanning operation. At first sight, it would seem logical to use arrays of transmitters and detectors with sequential activation and sampling of each transmitter/ detector pair. However, it has been found that the high cost of suitable devices precludes further investigation in this direction.

There are several requirements which are fundamental to the design of the scanning system. Firstly, the scanning beam must be normal to the image plane at all times. Secondly, the transmitted and reflected beams must be coaxial with one another. Thirdly, each sensor must scan a square, television like raster of 64 lines of 64 range samples per line. This must be done in approximately 2s, which means that range values are sampled approximately every 500ms.

A schematic diagram in plan view of one possible scanning system is shown in Fig. 5.

A normal scanning beam is produced by reflecting the laser beam off a plane mirror located at the focal point of a large convex or aspheric lens corrected for spherical abbreviation. Rotating the mirror in the plane of the diagram scans the beam horizontally across the image plane. An alternative method which is also under consideration uses a large parabolic mirror instead of a convex lens, again with the plane mirror at its focal point.

That part of the reflected radiation which is coaxial with the transmitted

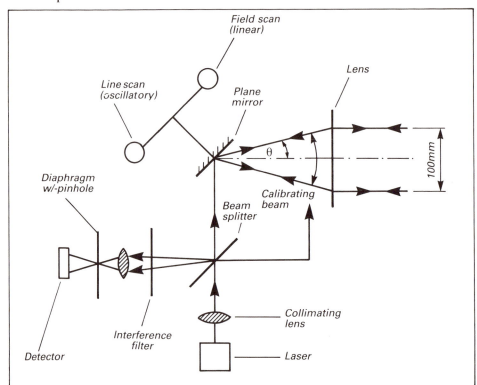

Fig. 5   Schematic of scanning system

beam returns along the same path, and a fraction of it is deflected by the beam splitter. It then passes through a narrowband interference filter with a passband of approximately 2nm, whose purpose is to reject radiation from all sources other than that particular sensor. In practice, the most troublesome sources of radiation are the other sensors. This problem is overcome by using a laser diode of slightly different wavelength for each sensor, each with its own interference filter. The wavelength of radiation from any laser need differ by only 10nm from that of any of the others to eliminate interference. The light output from the interference filter is then focused by a lens system onto a pinhole in a diaphragm. This limits the maximum off-axis deviation of the reflected light to less than 0.5° (depending on the size of the pinhole). The light then impinges on the surface of the avalanche photodiode.

Line scanning is achieved by oscillating the plane mirror by means of a proprietary optical scanner on which it is mounted. This is a small, precision moving-iron galvanometer device with a built-in capacitive transducer which permits accurate control of both mirror angle and angular velocity. Vertical, or field scanning is performed by incrementing the whole assembly shown in Fig. 5 downwards by approximately 1.5mm at the end of each line scan using a stepper motor drive operating in open-loop mode under computer control. Only a small section of the lens or parabolic mirror is then required. This method eliminates the pin-cushion distortion of the raster that would be inevitable (unless the lens or mirror were designed to compensate for it) if the mirror were tilted to produce the field scan.

A problem with systems of this type is the inherent variation of the path length and scanning velocity as the mirror rotates. In the ideal case, ignoring the effect of the lens, path length varies inversely with $(\cos \theta)$ where θ is the angle of the laser beam to the optical centre line of the lens. It is not a problem in this particular system because it is implicitly included in the compensation procedure for range values mentioned in the previous section. Effective scan velocity, however, varies inversely with $(\cos \theta)^2$ with constant angular velocity of the mirror during line scanning, and causes difficulties with obtaining range values equally spaced along each line of the raster. Two possible solutions are at present under consideration. The first, and less accurate method is to energise the optical scanner which rotates the mirror with a suitably shaped waveform driving into a closed-loop amplifier utilising the scanner's internal transducer. The second uses that part of the transmitted laser beam which is unavoidably reflected by the beam splitter (the 'calibration beam' in Fig. 5). This beam is redirected by a suitable arrangement of prisms and/or plane mirrors back into the line of the main beam, but displaced vertically by a small amount. The calibration beam strikes the plane mirror and is rotated through the same angle as the main beam. It then passes through a reticle of 64 equally spaced dark lines, and through a lens which focuses it onto a silicon photodiode–amplifier combination. A series of sampling pulses are therefore generated from the photodiode amplifier which are used to synchronise range sampling along each line of the scan. Thicker lines at each edge of the reticle can be used to generate synchronisation pulses for the vertical, or field scan mechanism.

Image processing algorithms

This section discusses techniques for analysing the range data obtained from the set of six sensors. The overall objective is to compute orientation information which can then be passed to a robot arm. The arm grasps the component, picks it up, reorientates it and places it at the workstation coordinates. There are, however, a number of subsidiary stages in attaining this objective. Firstly, it is necessary to ensure that the correct part is being examined and not some extraneous object which has escaped into the imaging station by mistake. Secondly, a degree of inspection is desirable; for instance, the component can be checked for absence of manufacturing defects. Thirdly, the orientation of the part with respect to that of a preprogrammed reference must be computed. Finally, the correct sequence of operations for reorientation must be specified and gripper application coordinates calculated.

As mentioned earlier, much work has been done concerning the use of television cameras and similar devices to image industrial piece parts in robot vision applications. Normally, an image from one viewpoint suffices and no attempt is made to extract range information. Simple, high-speed image-processing algorithms are used in order to achieve realistic throughput rates[1,2]. Most of the research on three-dimensional imaging, on the other hand, whether of industrial components or of more generalised subjects such as office or room scenes, tends to construct models based on, for example, combinations of flat and cylindrical surfaces[7]. The intention in this project is to try to find a compromise, namely simple high-speed algorithms designed expressly for three-dimensional engineering parts of complex shape. It is helped by several features of the application and the techniques under consideration which simplify the problems considerably. Firstly, the scene or image which must be analysed is a relatively simple one which consists, in the ideal case at least, of one component resting on a ground plane. Secondly, the use of range data obtained with an orthogonal scanning arrangement eliminates the effects of uneven illumination and perspective distortion. Thirdly, the imaging of the part from all sides reduces the proportion of missing data to a minimum.

A primary operation in many multiple-view systems, and one which can be very difficult, is the correlation of the separate views. The characteristics of this system, and also the exact edgewise registration of the six image planes one with another means that correlation is an inherent feature. The result of this is that if one point is 'seen' by two sensors, then the coordinates of the point for one sensor can be transformed to those for the other sensor by simply reordering the coordinates. Any point on the surface of a component, depending on its relative position on the component and the orientation of the latter with respect to the sensors, may be seen by none, one, two or three sensors. In general, therefore, there is always some degree of redundancy between the six views, a feature which can subsequently be put to use in eliminating quantisation error by averaging coordinate values of the same surface point which differ slightly.

A recognition process in which particular properties of the part under examination are compared with the corresponding ones of the preprogrammed component can be used both to check that the part is the right one and also to inspect it for defects. However, the former is essentially a rough preliminary test while the latter must be performed at a much higher resolution to pick out small defects. As the orientation of the types of component with which this investigation is concerned is often referenced with respect to very small features, inspection/final recognition is an operation which could usefully be combined with orientation computation.

Pertinent properties for a rough recognition test are comparisons of surface area and volume between the scrutinised and reference components. There is no need to ratio these two properties to obtain a shape descriptor as there is no requirement to classify objects of the same general shape but different sizes. Simple algorithms can be used to obtain rough measures of surface area and volume which will suffice for rejection of foreign bodies or grossly misshapen components.

Final recognition and orientation computation can be performed using a variation on procedures originally developed for similar operations on images of two-dimensional components viewed with a television camera[1,2]. In the original algorithms, a set of concentric circles are superimposed on the component outline held in software with their centres coincident with the centroid of the closed curve formed by the outline. Provided that the radii of the individual circles have been correctly selected to coincide with definitive features of the outline, the pattern formed by the intersections of the circles with the outline provides an orientation-invariant method of recognition and also a direct measure of orientation. In this investigation, the intersections between the surface of the component's three-dimensional 'range image' and a set of concentric spheres centred on the centre of gravity of the image are used for recognition and for orientation computation. The radii of individual spheres must be chosen to yield a unique pattern which can be recognised reliably during both preprogramming and operation.

The final operation which must be performed is the calculation of application coordinates for the robot-arm grippers. A fundamental requirement here is that the manipulator must grasp and pick up the component in a precise manner; that is, without disturbing its position or orientation significantly, otherwise the component will be incorrectly positioned when relocated. Gripping positions and angles must therefore be specified or computed as precisely as possible. A standard technique with two-dimensional components is the specification of pick-up points as offsets from some invariant feature such as the position of the centroid. Extrapolation to the three-dimensional case necessitates a set of pick-up points, one for each stable attitude of the part. From a flexibility point of view, it is preferable to compute gripping coordinates from a consideration of the uppermost surface of the component without relying on pre-programmed information. Research on the automatic handling of two-dimensional parts[9,10] has shown the latter to be a practical, though rather involved, proposition. Extrapolation to the three-dimensional case under discussion is more complex but not intractably so, as only that portion of the component seen in the top view need be considered in most cases. The advantages that accrue, as in the two-dimensional case, are reduction in preprogrammed information, higher probability of reliable handling, and an improvement in the ability of the system to cope with unforseen circumstances such as grossly misshapen or stuck-together parts.

Gripper system

This application makes particularly heavy demands on the design of suitable grippers for the robot arm performing the reorientation operation, because in many cases the part must be orientated in more than one plane. However, it involves the handling of one specific component type only at any one time. Foreign bodies or defective parts are detected by the imaging system and escaped from the imaging station down a chute. Grippers can therefore be developed for each particular component type.

The grippers must be able to pick up the part from any of the latter's stable postures. For many components this will necessitate two or more separate gripper configurations. A primary requirement is that the jaws of each gripper must be able to rotate ±180° in a vertical plane to turn parts upside down when necessary. This means in turn that to be able to deposit the component, the jaws must grasp it around its outer edges. The use of a vacuum sucker gripper must therefore be ruled out in general. The jaws must also be able to rotate in a horizontal plane, but for most applications the wrist rotation movement on the robot arm will suffice. The two other degrees of freedom of the robot arm wrist, namely wrist bend in two directions at right angles to one another, can be used to orientate the gripper assembly.

Actuation of the wrist joints and rotation of the gripper jaws can be performed by means of dc servomotors, with opening and closing of the jaws being pneumatically actuated. A major problem is the provision of two or more grippers on the hand which can be indexed into position as required. Multi-gripper indexing turrets are heavy and bulky and could limit the working envelope of the hand, although this latter problem can be alleviated to some extent by retracting the six sensors away from the imaging station during handling. A possible alternative is an automatic tool changer system for the grippers, with the robot arm selecting the correct gripper under program control.

Detailed gripper design obviously depends on component geometry and is difficult to predict. However, it is anticipated that two basic gripper configurations in various sizes will suffice for many parts. One consists of two flat, mutually opposed, parallel acting plates for gripping by parallel straight edges. The other consists of two notched jaws for gripping circular profiles or across corners. The jaws in each case are spring-loaded, and contact sensors are included to assure correct gripping of the part.

Concluding remarks

The range-finder imaging system described in this paper possesses several characteristics of importance to both automatic inspection in the wider context and to robot vision. It provides unambiguous precisely registered three-dimensional scene information from a number of view points. The data is free from illumination effects such as shading and shadows, does not suffer from perspective distortion, and contains errors due to reflections only when very highly polished surfaces are involved. The range-finder could form the basis of a very powerful robot system, with the number of individual sensors being specified according to the application.

The use of six identical sensors might be thought excessive, particularly as oblique surfaces produce a degree of overlap between the fields of view, which leads to some duplication of results. However, this can be advantageous. At large angles of incidence, the spot of light projected by the laser is elongated and an average range value is measured. This is exacerbated by the continuous line scanning mechanism. Although system parameters have been chosen so that the phase delay corresponding to the largest range in the image volume can be measured in less than 1% of the time between the start of successive range samples, nevertheless this compounds the error. Because of the large angle of incidence, however, one of the other sensors must be viewing the surface near normally, and can therefore provide more accurate range data which can be used in preference. The exception is where re-entrant surfaces such as the sides or bottoms of holes or depressions are being imaged at large angles of incidence and in such a way that none of the other sensors has an unobstructed view. In these cases, inaccuracies must be accepted.

The overall system is intended to orientate components which at present must be magazined, manufactured just prior to assembly, or coded with non-functional orientation features. It is not intended for simple parts which can be orientated by mechanical methods and fed at a rate of several hundred per minute. Estimates for operating time of the three main constituents of the system are as follows: imaging, 2s; data processing and scene analysis, 4s; handling and relocation 4s. It is expected that further development on the sensors could reduce total imaging time to approximately 0.5s. In addition, some overlapping of these operations will be possible in practice as a new inspection cycle can be started as soon as the robot arm has picked up the component. However, it is clear that these estimates represent a feed rate which is at least an order of magnitude slower than that for simple parts and mechanical orientation tooling. However, complex parts which cannot be orientated mechanically can be processed, and complete imaging is being performed which makes for high reliability.

References

[1] Pugh, A., Heginbotham, W.B. and Kitchin, P.W. 1972. Visual feedback applied to programmable assembly machines. In, *Proc. 2nd Int. Symp. on Industrial Robots*, pp. 77-88. IITRI, Chicago.
[2] Heginbotham, W.B., Gatehouse, D.W., Pugh, A., Kitchin, P.W. and Page, C.J. 1973. The Nottingham SIRCH assembly robot. In, *Proc. 1st Conf. on Industrial Robot Technology*, pp. 129-143. IFS (Publications) Ltd, Bedford, UK.
[3] Ward, M.R., Rossol, L. and Holland, S.W. 1979. 'Consight', a practical vision-based robot guidance system. In, *Proc. 9th Int. Symp. on Industrial Robots*, pp. 195-211. SME, Dearborn, MI, USA.

[4] Pugh, A., Heginbotham, W.B. and Waddon, K. 1978. Versatile parts feeding package incorporating sensory feedback. In, *Proc. 8th Int. Symp. on Industrial Robots*, pp. 206-217. IFS (Publications) Ltd, Bedford, UK.

[5] Ishii, M. and Nagata, T. 1976. Feature extraction of three-dimensional objects and visual processing in a hand-eye system using laser tracker. *Pattern Recognition (GB)*, 8: 229-237.

[6] Nitzan, D., Brain, A.E. and Duda, R.O. 1977. The measurement and use of registered reflectance and range data in scene analysis. *Proc. IEEE*, 65: 206-220.

[7] Duda, R.O., Nitzan, D. and Barrett, P. 1979. Use of range and reflectance data to find planar surface regions. *IEEE Trans. Patt. Anal. and Mach. Intell.*, PAM-1 (3): 259-271.

[8] Page, C.J., Pugh, A. and Heginbotham, W.B. 1976. New techniques for tactile imaging. *The Radio and Electronic Engineer*, 46 (11): 519-526.

[9] Page, C.J. 1974. Visual and tactile feedback for the automatic manipulation of engineering parts. PhD Thesis, University of Nottingham, UK.

[10] Page, C.J. and Pugh, A. 1980. Visually interactive gripping of engineering parts. Infotech state-of-the-art report on the Factory of the Future.

5
Structured Light/Scene Illumination

Subtle use of light (including structured light) is the secret to successful robot vision. This topic cannot be divorced from 'the sensor' and is included here to complete the picture.

SCENE ILLUMINATION

D.A. Hill
CUEL, UK

Information on the static and dynamic characteristics of the other
elements in robot vision and sensing scenes is the starting point for
specifying scene illumination. That information enables the most suitable
illumination hardware to be identified and evaluated thereby facilitating
system commissioning and installation.

There are several respects in which illumination for robot vision and sensing
based on light differs from that for scenes which do not have robots as one of
their elements. In the case of scenes involving robots, the main feature is that
the majority, if not all, the five elements – object, sensor, surroundings,
robot and the sources of illumination – may be moving. The movement may
be either continuous or intermittent and there may be a further complicating
factor for designers of scene illumination systems.

The movement involves changes of direction and amplitudes only
experienced when working with robots, so little previous experience is
available for reference.

The number of movements in total, their direction, magnitude/amplitude
and particularly their sequence, present problems for interfacing illumination with the other four elements in the scene. Fortunately, designing the
illumination system is not totally dependent on laboratory or on site
experimentation. Basic data, both optical and mechanical can be collected
for all the five elements in the scene from manufacturers' catalogues,
engineering drawings and the scene itself.

Correlating the data for the different elements in what is essentially a
dynamic scene presents considerable problems. Producing a series of
annotated isometric sketches – each sketch 'freezing' a different stage of the
scene – greatly facilitates conceptualisation and modelling. If the facilities
were available CAD would obviously replace this more laborious method.
An illumination specification can then be written as a prerequisite for
procuring the necessary hardware.

Installing the hardware will, in all probability, reveal the need for
adjustments and replacements, but if this ground work is thorough it will
assist greatly with installation and commissioning. The expeditious isolation
and tracing of any illumination problems will be the principal advantage.

As all the elements in the scene are interconnected, consideration is given first to the context of illumination for robot sensing. The other four elements in the scene are therefore considered and related to illumination before embarking on a fuller consideration of illumination in its own right.

The object

The dimensions of the object, its shape and, in the case of assemblies and subassemblies, constituent parts may be taken from engineering drawings and then confirmed or even acquired by metrological equipment. Data on manufacturing as opposed to drawing tolerances must be used. When all the manufacturing tolerances for all the elements in a scene are aggregated, deviations from the nominal scene may be significant. The illumination has to be capable of accommodating these deviations so this data is vital. At a later stage the dimension data has to be translated into optical engineering terms, e.g. beam angles for illumination.

Engineering drawings often record non-dimensional information. This information forms the basis of the optical data which is needed for designing a scene illumination system. Information on materials and surface finish taken from drawings facilitates comparison with the average values for typical materials and finishes listed in reference tables.

However, the interaction of material, surface-finish and shape, make a consideration of an object's micro-topography essential. The importance of this increases with the amount of magnification needed by the sensor.

Average data provides only general indications. Micro-topography helps avoid minute but serious impediments, e.g. flare spots and black holes. Both impediments in their different ways prevent information reaching the sensor and thereby make the system inoperative.

Opaque objects are best understood as agglomerations of microscopic compound mirrors with either hard or soft edges. Each mirror has its own characteristics which may conflict with those of its neighbour. These affect the direction and amount of received light reflected. The surfaces of transparent objects also act as mirrors although refraction and diffraction are possibly more important optical phenomena.

Opaque, transparent and translucent objects all selectively absorb light in varying amounts. The absorption is not uniform across the spectrum. If the absorption by the object is greatest at those wavelengths where the sensor is least able to compensate the scene, illumination of greater intensity in the right places is a must.

The surroundings

The surroundings in which the object sensor and robot operate are the second element in the scene. If there is insufficient contrast between an object and its surroundings the sensor's performance will be impeded. Difficulty will be experienced in distinguishing the outer edges of the object. The robot's positional accuracy and repeatability will consequently be adversely affected. Changing the direction of illumination affords some improvement in edge definition. The presence of ancilliary equipment in the surroundings, static and moving, also imposes constraints. It affects the space available and the choice of positions for mounting sources of illumination. Working distances may be affected by these obstacles. These in turn affect beam angles, lamp dimensions, illumination intensity and the ancillary optical components needed for overcoming the constraints. On the one hand, the obstacles may necessitate long working distances; yet on the other hand, they may need the illumination equipment to work in close proximity to the object.

The sensor

Manufacturers of electro-optic sensors, even simple single-element ones, invariably provide comprehensive product data. This data is invaluable for interfacing sensors with illumination. It has, however, to be related to data for any optical components in systems associated with the sensor.

These components, e.g. lenses, have their own transmission characteristics, which almost always differ from those of the sensor. In addition, these components determine the field of view of the sensor and its viewing direction. As it is not always the practice to specify lenses in this way, some optical bench or laboratory experiments may be necessary to acquire this data. It is especially important to identify any tolerances applicable to those specifications, e.g. if an iris is used with a lens, to what extent does it change the field of view?

The robot

The robot is the fourth element in the scene. All robots have two characteristics which impose further constraints on illumination. First the lifting capacity available for illumination equipment may preclude the use of attached illumination. Detached illumination is then mandatory. The weight of low-voltage transformers, tungsten halogen lamps and steel enclosures may exceed the lifting capacity available.

Secondly, movements of the robot arm may likewise introduce further problems affecting illumination. There are lamps which should only be used in certain burning positions. A transient movement through one of the forbidden burning positions may not be detrimental to the performance of the lamp. Remaining stationary in that position would however be detrimental.

When the illumination is attached, vibration associated with the movement, acceleration and deceleration of the robot arm could affect the life and performance of lamps if unsuitable ones are selected. Lamps made for automotive use – buses, cars and lorries – are the ones most resistant to the effects of vibration. There are also tungsten filament (incandescent) lamps which are sold for 'rough service'.

Vibration can cause the filament to chatter or bounce if the distance between supports is too great and if the diameter of the tungsten wire is too fine. The sensor then has to contend with a fluctuating field of illumination. Characteristics such as this are confirmed by experiment when the illumination system is installed and may even necessitate using a different lamp as a source of illumination or antivibration mountings.

Positional tolerances on a robot's movements may be large enough to make the choice of homogeneous illumination system mandatory. Irregularities, e.g. 'islands' of shadow within the field of illumination, could adversely affect the performance of the sensor. These irregularities by virtue of being intermittently superimposed on the target could adversely affect the repeatability performance of the sensor. Homogeneity would safeguard this. There are occasions when a function of an illumination system is to project a grid from a graticule onto the subject. The illumination has to be sufficiently homogeneous to avoid any loss of grid definition due to the interference of amorphous shadows attributable to components inside a lamp.

The illumination systems

This comprises lamps, ancilliary electrical and mechanical equipment, and optical accessories. There are no lamps which are specially made for use in conjunction with robot sensors. Lamps have to be chosen from those made for other purposes. Those made for automotive use have already been mentioned previously.

In all, there are five categories of lamps which are potential candidates for robot sensor use. Each category includes more than one type of lamp. These broadly coincide with sections in lamp manufacturers' catalogues.

If the automotive lamps, tungsten filament and tungsten halogen, are regarded as the first category, the second one comprises lamps for domestic, shop and office lighting. The lamps are either tungsten filament or fluorescent, and operate at all mains voltage. They are not always available in 110V versions. If tungsten filament lamps are attached to the robot, care has to be exercised in choosing a lead with sufficient flexibility and insulation. Low wattage fluorescent lamps for caravans are also in this category.

The weight of the associated control gear and other components, which together make up a luminaire, favours the use of fluorescent lamps rigidly mounted in one position and continuously burning as the most usual configuration. Their slow start response and the adverse effect on their life of switching exclude the possibility of intermittent firing of a bank of these lamps fixed at various points in an illumination scene.

Although the majority of fluorescent lamps are linear, a small number of circular ones are available. Both types are useful for 'flooding' the scene.

The third category is that made up of tungsten halogen lamps for slide projectors. Compact source iodide and xenon lamps for cinema projection are also in this category, but they are not considered suitable for use with robot sensors. With the exception of the lower wattage lamps up to 50W, forced air cooling is needed.

As spot sources they are more compact, versatile, intense and efficacious than their nearest tungsten filament counterpart. They interface fairly easily with optical components. Some are available with an integral dichroic reflector which eliminates the need for a separate cold light mirror or reflector.

The fourth category is instrument lamps. They are either miniature tungsten filament or tungsten halogen. The latter are not widely available.

Table 1   Classification of lamp data

Electrical input	Output	Anatomy	Ancillary equipment	Constraint	Compatibility with:
	Intensity – lumens – MCDs (LEDS)				

Voltage
– normal
– underrun
Wattage
Amperage –
  miniature
  lamps
	Spectral
	distribution
		Envelope
		– dimensions
		– clear/pearl
		– shape
			Transformer
			Fan
			Control gear
			Reflector
			Enclosure
				Burning
				position
				
				Life
					Lenses
					– spherical
					– cylindrical
					– glass/acrylic

	Area spread of luminous flux	Filament – position – shape – rigidity	Electrical lead Electrical stabiliser – rectifier – compensator	Thermal limits	Prisms Diffusers Mirrors Filters Graticules Condensers Diaphragms
	Bright and dark 'islands' (isophots)				
	Rise and fall times after switching		Homogeneous		Windows – sapphire – quartz – hard glass

Both types are available with and without integral lenses. The lenses are not high precision, so the remarks already made elsewhere on the subject of tolerances are relevant here. They are small enough to be grouped together in arrays and banks but they do need a low voltage electrical supply. Convectional cooling is adequate for both types.

The fifth category of sources of illumination for robot sensors merits consideration on account of the increased sensitivity of sensors. Visible, usually red-light-emitting diodes of high brightness, often provide sufficient illumination for robot sensors. Two configurations are used: cylindrical with an integral lens, and stackable which give a line of light. In some cases they have substituted for fibre-optic cross-section converters.

Lamps for streets, sports grounds and factory high-bay lighting are not considered suitable for use with robot sensors. They are large in physical size and output. If 'floods' of light are needed on the scene there are lamps in the first three categories which adequately serve the purpose.

Further analysis of the five categories is needed to finalise the design of an illumination system. Table 1 lists the headings.

Proprietary luminaries and fibre-optic light sources consolidate most of the above information for the designer. Thus luminaires are described in terms of their beam angles, overall dimensions and working distances. In the unlikely event of decorative luminaires being suitable for use with robot sensors much of the information has to be acquired by experiment because they are made primarily for aesthetic purposes.

Fibre-optic light sources constitute a second but small group of specialist luminaires. Besides average lamp life the published data typically lists the weight and dimensions of the light source, the rating of its tungsten halogen lamps and the diameter of the optical fibre bundle which it can supply. On the basis of that data, the scope for using single and multi-arm light guides, focused and unfocused, and rectangular slits may be calculated. Provided

Table 2   Scene illumination specification

Mechanical	Optical	Electrical	Environmental
Mounting position(s) – attached – detached and location	Intensity (intensities)	Lamp rating(s) Burning – continous	Cooling – convection – forced air
Weight(s) (attached only)	Shape(s) and dimensions – solid circle(s) – rectangle(s)	– intermittent with frequency, sequence and duration for each	Cleaning – continuous air purge – manual
Overall dimensions	– annulus (annuli)	lamp	
Arrangement of lamps – single spot – multiple array/bank	Evenness – homogeneous – amorphous 'islands'	Number of lamp(s) Replacement of consumable items	
	Working distance(s)		
	Spectral distribution(s)		
	Direction(s)		
	Associated optical component(s)		

NB, Brackets have been used with those items which may be specified either singly or in multiples

they can withstand the degree of flexing, fibre-optics is an effective compromise between detached and attached illumination. The light source may be detached whilst the optical fibre component is attached.

For the majority of robot sensor applications, data has to be consolidated by the system designer. Lamps, electrical and mechanical accessories and optical components all have to be correlated. Table 2 lists the information which has ultimately to be acquired by consolidating the data relating to the illumination components.

The specification based on Table 2 will almost certainly go through several earlier editions. The first will correspond to the information obtained by modelling the system. The last edition will record data on the scene illumination as confirmed by installation, commissioning and testing.

CONSIGHT-I: A VISION-CONTROLLED ROBOT SYSTEM FOR TRANSFERRING PARTS FROM BELT CONVEYORS

S. W. Holland, L. Rossol and M. R. Ward
General Motors Research Laboratories, USA

CONSIGHT-I is a vision-based robot system that picks up parts randomly placed on a moving conveyor belt. The vision subsystem, operating in a visually noisy environment of manufacturing plants, determines the position and orientation of parts on the belt. The robot tracks the parts and transfers them to a predetermined location. CONSIGHT-I systems are easily retrainable for a wide class of complex curved parts and are being developed for production plant use.

In many manufacturing activities parts arrive at workstations by means of systems that do not control part position. Since present robots require parts to be in precisely fixed positions, their use is precluded at these workstations. To automate these part handling operations, intricate feeding devices that precisely position the parts are required. Such devices, however, are often uneconomical and unreliable.

Robot systems equipped with vision represent an alternative solution. This paper describes CONSIGHT-I, a vision-based robot system for transferring unorientated parts from a belt-conveyor to a predetermined location. CONSIGHT-I:

- Determines the position and orientation of a wide class of manufactured parts including complex curved objects.
- Provides easy reprogrammability by insertion of new part data.
- Works on visually noisy picture data typical of many plant environments.

As a result of these characteristics – and because the vision subsystem does not require light tables, fluorescent conveyor belts, coloured parts or other impractical means for enhancing contrast – CONSIGHT-I systems are eminently suitable for production plant use.

Functional overview

CONSIGHT-I functions in two modes: a set-up mode and an operational mode. During set-up, various hardware components are calibrated and the system is programmed to handle new parts. Once calibrated and programmed for a specific part, the system can be switched to operational mode to perform part transfer functions.

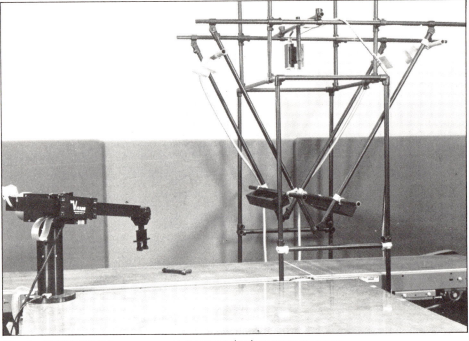

Fig. 1 CONSIGHT-I conveyor, camera, and robot arrangement

Part transfer operates as follows: operators place parts, such as foundry castings, in random positions on the moving conveyor belt (Fig. 1). The conveyor carries the parts past a vision station which is continually scanning the belt. Position and orientation information is sent to a robot system. As the parts continue to move, the robot tracks the moving parts on the belt, picks them up, and transfers them to a predetermined location. The above sequence operates continuously with no manual intervention except for placing parts on the conveyor.

CONSIGHT-I is capable of handling a continuous stream of parts on the belt, so long as these parts are not touching. The maximum speed limitation is imposed by neither the vision system nor the computer control, but by the cycle time of the robot arm.

CONSIGHT-I is easily reprogrammed for a new part. The new part to be picked up is passed by the vision station. The vision subsystem determines part location relative to the belt. The belt is stopped and the robot hand is manually positioned at the desired pick-up point for that part. The set-up programs then automatically determine the transformation between the location determined by the vision station and the pick-up point used for grasping this part, and the system is ready to handle the new part.

CONSIGHT-I must also be calibrated when it is first set up, or whenever the camera, the robot or the conveyor is moved. The calibration procedure is simple and requires about 15 min. All necessary mathematical transformations are derived automatically and are totally transparent to the operator of the system, that is, he need not understand nor even be aware of the mathematical processes involved.

System overview

Organisation

CONSIGHT-I is logically partitioned into independent vision, robot, and monitor modules, permitting these system functions to be distributed among several smaller computers. Although the experimental system described here was implemented on a single computer, communication between subsystems was designed to allow easy substitution of new vision modules or new robots. The vision subsystem reports only a unique point (i.e. x and y coordinates) for each part, and an orientation. Neither the location of the point on the part, nor the reference from which to measure orientation, is specified. Our particular vision subsystem reports the part's centre of projected area and a direction along the axis of minimum (or maximum) moment of inertia. Other vision modules employing different vision techniques are available which report an altogether different point and orientation for the same part [1,2]. These other vision modules can easily be substituted for the first without changing the control program, the robot system, the part-programming methods, or the calibration methods. More importantly, it is equally easy to substitute another robot or a different monitor subsystem to perform a new and entirely different operation on the part.

Hardware

The hardware for the experimental version of CONSIGHT-I as described in this paper is shown in Fig. 2. The computer is a Digital Equipment Corporation PDP 11/45 operating under the RSX-11D real-time executive. The camera is a Reticon RL256C 256 × 1 line camera. The robot is the Stanford Arm made by Vicarm [3].

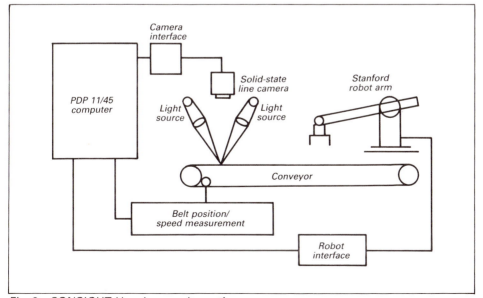

Fig. 2   CONSIGHT-I hardware schematic

Since its speed is neither constant nor predictable, the belt is instrumented with a position and speed detection device. Position and speed information are necessary for three reasons:

- The camera scans the belt at a constant rate, independent of belt speed. For each equal increment of belt travel, the vision subsystem records one of these scans. Belt travel increments must therefore be measured precisely.
- Programs must compensate for movement of the belt between picture digitisation and part pick-up time. This also requires belt position detection.
- The robot needs both position and velocity information in order to track the moving part on the belt smoothly.

The specific implementation details for production versions of CONSIGHT-I will differ from those described here. The functions handled by a single computer will be distributed among three smaller computers communicating with each other and the robot will be replaced with one suitable to a production environment.

Software

The software organisation for CONSIGHT-I reflects the three major modules of the system. The monitor coordinates and controls the operation of CONSIGHT-I and also assists in calibration and reprogramming for new parts. The monitor queues part data and the system is thus capable of handling a continuous stream of parts on the belt.

The vision subsystem uses a modified range-finder approach in which two projected light lines, focused as one line on the belt, are displaced by objects on the belt. The line camera, focused on the line, detects the silhouette of passing objects. When it has seen the entire object, the vision subsystem sends to the monitor the object's position and a belt position reference value.

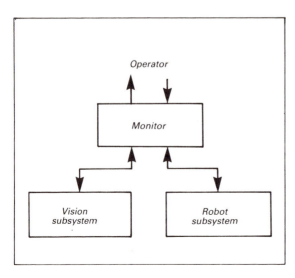

Fig. 3   CONSIGHT-I software organisation

Since the vision subsystem detects only silhouettes, parts must have silhouettes that allow unique orientation determination, or they must be rotationally invariant (i.e. part orientation cannot be defined or is unimportant). The idea of using structured lighting to simplify visual processing may also be found in other research [4–6].

The robot subsystem executes a previously 'taught' robot program to transfer the part from the conveyor to a fixed position. It accepts information concerning the part's location on the moving belt and uses this data to update the 'taught' program. It then monitors belt position and speed to track the part along the moving belt, pick up the part and transport it to a predetermined location.

Vision subsystem

The vision subsystem detects parts passing through its field of view and reports their position and orientation to the monitor program. Parts may follow in an unending stream. It is also permissable for several parts to be within the field of view simultaneously. Parts which are overlapping or touching each other are ignored and allowed to pass by the robot for subsequent recycling.

The vision subsystem employs a linear array camera. The linear array images a narrow strip across the belt perpendicular to the belt's direction of motion. Since the belt is moving, it is possible to build a conventional two-dimensional image of passing parts by collecting a sequence of these image strips. The linear array consists of 256 discrete diodes, of which 128 are used in the system described here. Uniform spacing is achieved between sample points (both across and down the belt) by use of the belt position detector which signals the computer at the appropriate times to record the camera scans of the belt.

The two main functions of the vision subsystem are object detection and position determination.

Object detection

A fundamental problem which must be addressed by computer vision systems is the isolation of objects from their background. If the image exhibits high contrast, such as would be the case for black objects on a white background, the problem is handled by simple thresholding. Unfortunately, natural and industrial environments seldom exhibit these characteristics. For example, foundry castings blend extremely well with their background when placed on a conveyor belt. Previous approaches for introducing the needed contrast, such as the use of fluorescent painted belts or light tables [7], are impractical. They would severely restrict the number of potentially useful applications of vision-based robot systems. We developed a unique lighting arrangement which accomplishes the same result without imposing unreasonable constraints on the working environment.

The principle of the lighting apparatus is illustrated in Fig. 4. A narrow and intense line of light is projected across the belt surface. The line camera is positioned so as to image the target line across the belt. When an object passes into the beam, it intercepts the light before it reaches the belt surface.

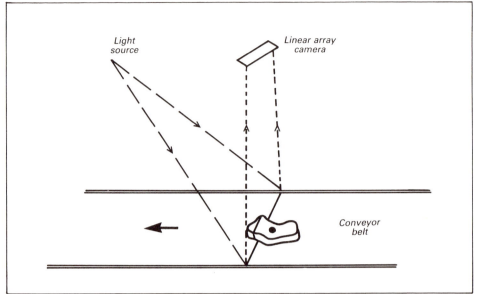

Fig. 4   Basic lighting principle

When viewed from above, the line appears deflected from its target wherever a part is passing on the belt. Therefore, wherever the camera sees brightness, it is viewing the unobstructed belt surface; wherever the camera sees darkness, it is viewing the passing part (Fig. 5).

Unfortunately, a shadowing effect causes the object to block the light before it actually reaches the imaged line. The solution is to use two (or more) light sources all directed at the same strip across the belt. Fig. 6 illustrates the idea. When the first light source is prematurely interrupted, the second normally will not be. By using multiple light sources and by adjusting the angle of incidence appropriately, the problem is essentially eliminated.

The light source is a slender tungsten filament bulb focused to a line with a cylindrical lens. Fig. 7 illustrates the line of light generation hardware.

The described lighting arrangement produces a height detection system. Any part with significant thickness will be 'seen' as a dark object on a bright background. The computer's view is a silhouette of the object. Note that while the external boundary will appear sharp, some internal features, such

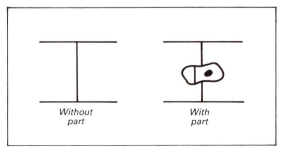

Fig. 5   Computer's view of parts

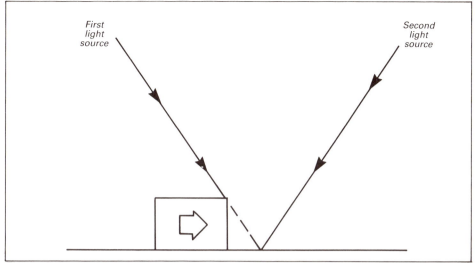

Fig. 6   Improved lighting arrangement

as holes, are still subject to distortion or occlusion due to the shadowing effect. These internal features may optionally be ignored or recorded under software control.

It is the computer's responsibility to keep track of objects passing through the vision system. Since several pieces of an object or even different objects may be beneath the camera at any one time, the continuity from line to line of input must be monitored. The conventional image segmentation schemes are the four-connected and the eight-connected algorithms. Both of these algorithms result in ambiguous situations[8] when deciding on the

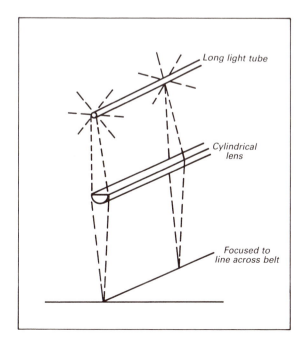

Fig. 7   Line of light generation

inside/outside relationship between some binary regions. A clever solution to that problem is the use of six-connected regions[9]; that is, connectivity is permitted along the four sides of a picture element and along one of the diagonals. At the expense of a minor directional bias in connectivity determination, the inside/outside ambiguity is resolved. In addition, the algorithms which implement the segmentation for six-connected regions remain simple and symmetric with respect to black and white. The six-connected algorithm artificially introduces the pleasing properties gained through hexagonal tessalation.

The six-connected binary segmentation algorithm is readily adapted for run-length coded input; that is, where only the transition points between black and white segments are recorded. This is a significant advantage. The straightforward binary segmentation algorithm requires that the intensity of the neighbours for each pixel be examined. The execution time is therefore 'order n squared' where n is the linear camera resolution. Since the number of black/white transitions across a line is relatively independent of the resolution for these types of images, the execution time is reduced to 'order n' for the algorithm using the run-length coding scheme.

Once the passing objects have been isolated from the background, they may be analysed to determine their position and orientation relative to a reference coordinate system.

Position determination

For each object detected, a small number of numerical descriptors is extracted. Some of these descriptors are used for part classification; others are used for position determination.

For position specification, we describe the part's position by the triple (x, y,θ). The x and y values are always selected as the centre of area of the part silhouette. For most parts, this represents a well-defined point on the part.

There is no convenient method for uniquely assigning a θ value to all parts. However, one useful descriptor for many parts is the axis of the least moment of inertia (of the part silhouette). For long, thin parts, this can be calculated accurately. The axis must still be given a sense (i.e. a direction) to make it unique. This is accomplished in a variety of ways and is part specific. The internal computer model for the part specifies the manner in which the θ value should be computed. For example, one method available for giving a sense to the axis value is to select the moment axis direction which points nearest to the maximum radius point measured from the centroid to the boundary. Another technique uses the centre of the largest internal feature (e.g. a hole) to give direction to the axis. Several other techniques are also available.

Parts which have multiple stable positions require multiple models. Parts whose silhouettes do not uniquely determine their position cannot be handled.

Reprogramming the vision system for a new part requires entering a description of a new model. Each model description includes information to determine if a detected object belongs to the class defined by the model and also prescribes how the orientation is to be determined.

The vision system sees the world through a narrow slit. As objects pass through the slit, statistics concerning that object are continuously updated. Once these statistics have been updated, the image line is no longer required. Consequently, storage need only be allocated for a single line of the two-dimensional image, offering a major reduction in memory requirements.

The block of statistics describing an object in the field of view is referred to as a component descriptor. The component descriptor records information for every picture element which belongs to that component. This includes the following:

- External position reference.
- Colour (black or white).
- Count of pixels.
- Sum of x-coordinates.
- Sum of y-coordinates.
- Sum of product of x- and y-coordinates.
- Sum of x-coordinates squared.
- Sum of y-coordinates squared.
- Min x-coordinate and associated y-coordinate.
- Max x-coordinate and associated y-coordinate.
- Min y-coordinate and associated x-coordinate.
- Max y-coordinate and associated x-coordinate.
- Area of largest hole.
- x-coordinate for centroid of largest hole.
- y-coordinate for centroid of largest hole.
- An error flag.

Considerable book-keeping is required to gather the appropriate statistics for passing objects and to keep multiple objects segregated. The primary data structure used for this purpose is the 'active line'. The active line records the location of each black and white segment beneath the camera and also the objects to which they are connected. For every new segment which extends a previously detected object, the statistics are simply updated. If the segment is the start of a new object which is not in the active line, it will be added to the active line and have a new component descriptor initialised for it. When an object in the active line is not continued by at least one segment in the new input line, it must have passed completely through the field of view. The block of statistics is then complete and may be used to identify the object and compute its position and orientation.

Fig. 8 illustrates the coordinate reference frames used by the programs. It is convenient to consider the x-origin of the vision system to be permanently attached to the belt surface moving to the left. Normally, this would cause the current x-position beneath the camera to climb toward infinity. To avoid this, all x distances are measured relative to the first point of object detection. This, in turn, induces a complication.

The complication occurs when two appendages of one object begin as two separate objects in the developing image. It then becomes necessary to combine the component descriptors for the two appendages into a single

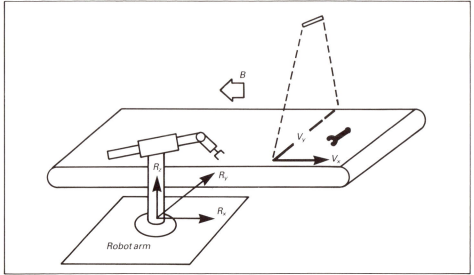

Fig. 8   CONSIGHT-I coordinate systems

component descriptor which reflects the combination of the two. Moments, however, have been referenced to two different coordinate systems (i.e. the *x*-origin for each was taken as the point of first detection). The required shifting of moments is accomplished by applying the Parallel Axis Theorem.

Once parts have passed completely through the field of view, final position determination is made. These computations proceed asynchronously with respect to the processing of the new lines of picture data. The results must be provided to the monitor subsystem before the part travels past the robot's pick-up window. To coordinate the scheduling of these final computations, a queue of completed component descriptors is maintained. Component descriptors are removed from the head of the queue and processed as time permits.

Since the belt on which the parts rest is moving, the vision system records the current belt position whenever a new object appears in the developing image. This belt position reference value is obtained from the belt position/speed decoder. Since the leading edge of each part defines the origin of the coordinate system to be used for that part, the position of that part at some future time can then be readily determined by checking the current belt position and adjusting for any belt travel since the initial reference was recorded.

Robot subsystem

The robot programming subsystem is implemented as two independent tasks. One task is required for robot program development and is necessary only during the programming and teaching phase. The second task, the run-time control system, is required both during the teaching phase and during robot program execution. It interprets and executes a robot program and controls the robot hardware. The execution of this task is controlled by special requests sent from other tasks.

A robot program consists of statements specifying: a position to which the robot should move (setpoint), an operation the robot should perform, or the environment for subsequent execution. Positions to which the robot moves are either taught by moving the robot manually and recording the position or are programmed by entering the specific Cartesian coordinates of a point in space from the keyboard.

In addition to this basic programming support, tracking and real-time program modification were developed for CONSIGHT-I. Tracking provides the ability to execute a robot program relative to some moving frame of reference. Program modification provides the ability to modify, in real-time, the robot program under external program control and thereby dynamically modify the robot's path.

Tracking is implemented by defining new reference coordinate systems called FRAMEs[10]. Normally the robot operates in a Cartesian coordinate system [R] with its origin at the base of the robot (Frame 0). The robot's Cartesian position is described by a matrix [P] which defines the position and orientation of the hand in [R]. The arm solution program then determines a joint vector [J] from [P].

$$[P] \longrightarrow [J]$$

If, however, we want to define [P] relative to a different coordinate system (frame) whose position in [R] is defined by a transformation [F], then the solution program must perform the following:

$$[F]\,[P] \longrightarrow [J]$$

Frames provide a means of redefining the frame of reference in which the robot operates. The robot may be programmed relative to one frame of reference and executed relative to a different frame of reference. For example, a robot may be programmed to load and unload a testing machine. If the testing machine is moved, the entire program can be updated by simply redefining the frame specifying the position of the testing machine without reprogramming each individual position point in the program. The overall effect is to translate/rotate every position to which the robot moves.

In addition to having a position, a frame is defined with a velocity and a time reference. This position, velocity, and time reference are used to compute or predict the frame's position. Each time the run-time system performs an arm solution (i.e. transforming the position matrix into the corresponding joint angles) it first computes a predicted position for the current frame.

Program modifications are special asynchronous requests sent to the run-time system from other tasks. Via these requests, an external program may modify a robot's programmed path, read the robot's position, start/stop robot program execution, and interrogate status – all while the robot is operating. These requests provide a means for greatly expanding the capability of the robot system without a major effort in developing a powerful robot programming language. Much of the logical, computational, and input/output capability of an algorithmic language (Fortran) is available for programming and controlling the robot external to the normal robot programming and control system.

In CONSIGHT-I, the part position determined by the vision subsystem defines the position and orientation of a frame. The approach, pick-up, and departure points are all programmed relative to this frame. The frame is also assigned the velocity of the belt. The robot subsystem does not directly interface to the belt encoder for belt position and velocity data, but receives the data via a request in the same way that the vision data is furnished. Thus, the rate at which the belt position and velocity data are updated is controlled by the monitor program and is a function of the variability of the belt speed. The approach, pick-up, and departure points are dependent upon the type of part being picked up as well as its position. Thus, these three programmed points are modified for each cycle of the robot.

System monitor

The monitor coordinates the operation of the vision and robot subsystems during calibration and part programming as well as during the operation phase. As stated earlier, calibration is required during the initial set-up or whenever the camera, the robot or the conveyor have been moved. Part programming is required only when modifying the system to handle a new part.

Calibration

Calibration is the process whereby the relationship between the vision coordinate system [V] and the robot coordinate system [R] is determined (Fig. 8). In particular we want to compute the position of a part [r] in [R] given the part position [v] in [V]. Taking into account belt travel, this computation is represented by the following equation:

$$[r] = [T] [v] + s \, b \, [B]$$

In the equation above, [T] is the transformation between [v] and [R], s is a scale factor relating belt distance to robot distance, b is the distance of belt travel, and [B] is the belt direction vector relative to [R]. Thus, [v] and b are the independent variables, and [T], s, and [B] are the unknowns to be determined by the calibration procedure.

To determine [B] and s, a calibration object is placed on the belt within reach of the robot hand. The hand is manually centred over the calibration part, the hand position is read, and a belt encoder reading is taken. The conveyor belt is started and the part is allowed to move down the belt. The robot hand is again centred over the calibration part. A second hand position and belt encoder reading are taken. The monitor system can now compute the belt direction vector [B] and a scale factor s, converting belt encoder units to centimetres.

Determining the coordinate transformation between [V] and [R] completes the calibration. The procedure assumes that the plane [Vx, Vy] is parallel to the plane [Rx, Ry] and that scaling is the same in both the x and the y directions.

A calibration object is again placed on the belt and allowed to pass by the vision station. The object position [v1] is determined by vision and a belt reading is taken. The part moves within reach of the robot and the conveyor is stopped. The robot hand is centred over the calibration object, the robot position [r1] is read and the distance of belt travel b1 is computed. This procedure is repeated with the calibration object placed at a different position on the belt resulting in a second set of data, [v2], b2, and [r2]. Combining these two sets of data points into the form above yields:

$$[r1 \; r2] = [T] \, [v1 \; v2] + s \, [B] \, [b1, \, b2]$$

This gives us four equations for determining the four unknowns in the transformation matrix [T].

Part programming

Part programming is the procedure for defining the gripper position for part pick-up. To do this, the vision subsystem must have previously been programmed to recognise the new part as described above. Generally the pick-up position [p] is offset from the part position as determined by vision. Thus, once the vision subsystem locates a part and its position [r] is computed as described in the preceding section, the actual robot pick-up position [p] must still be computed.

Fig. 9 illustrates this problem in a general way. The part position [v] and its rotation θ have been determined by the vision subsystem in the coordinate system [V]. The robot operates in [R] and needs to know the pick-up position [p]. Since the part may have any orientation on the belt, [p] is a function of both [v] and θ. [V] is rotated and translated from [R] as determined by [T] above.

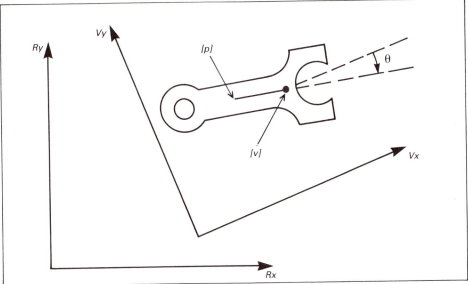

Fig. 9   The pick-up problem

The procedure is to pass a part by the vision station where its position and orientation are determined and a belt reading is taken. The part moves down the belt to within reach of the robot and the belt is stopped. The monitor subsystem computes the current position of the part in the robot's coordinate system [R] and sends this position and the part direction (as determined by θ) to the robot subsystem as the definition of a frame. The robot hand is then placed on the part in the desired position for the grasp and pick-up. This hand position is read by the monitor, but the robot subsystem returns the hand position relative to the newly defined frame, not the base coordinate system of the robot. Thus, the pick-up point is determined relative to the part position and oritentation as determined by the vision subsystem.

Later, during the operation phase, the part's position determined by the vision subsystem is again computed relative to [R]. This position and the part orientation defines a frame in which the robot operates during the pick-up phase of the robot program. The pick-up position is then simply the position determined by the procedure above. The robot subsystem automatically determines the actual robot position from the frame position, the pick-up point, and the belt position and velocity.

The operation phase

The final function of the monitor subsystem is to control the operation of the CONSIGHT-I system. Fig. 10 illustrates the overall logic of this portion of the monitor. Although the logic is straightforward, the monitor system deals with two rather subtle problems.

The first involves modification of the part pick-up point. The robot has a limited range of motion in all of its joints. In particular, the outermost joint, which controls hand orientation when the hand is pointing down, has only 330° of rotation. Clearly, this limited rotation does not permit every possible hand orientation. Since the gripper is symmetric about 180° we can effectively achieve all rotations by rotating the outer joint by ± 180° for some orientations. The monitor system determines when the orientation needs to be changed for a particular pick-up position. If the pick-up is modified, the put-down orientation also has to be similarly modified so that all parts are put down with the same orientation.

This orientation problem arises whenever the robot path is dynamically defined. The general problem, for grippers or for parts that are not symmetric or when the gripper is not pointing straight down, is not solvable with commercially available robots. Robots that provide wrists with greater flexibility (greater joint range) are required.

Error recovery is the second problem. The possibilities for parts out of reach, impossible robot positions, collisions, etc. increase significantly when the robot's path is being changed for each cycle. Errors detected by either the vision or the robot subsystems are reported to the monitor. The monitor then takes action, usually restarting the robot. Like the orientation problem above, however, the dynamic nature of the system makes a good general solution to error handling difficult. Further research and experience with installed systems is needed to determine how to best approach the error handling problem.

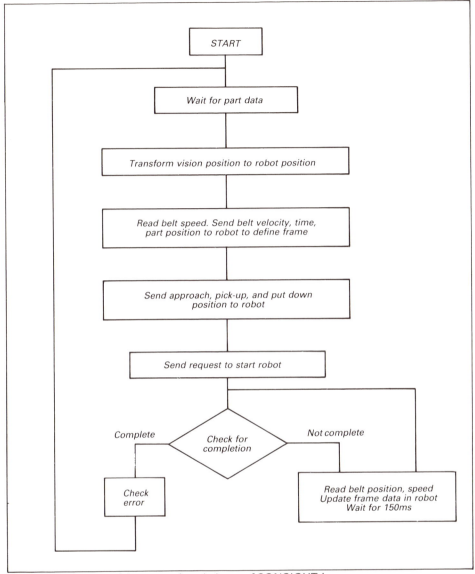

Fig. 10   Overview of the operational phase of CONSIGHT-I

Concluding remarks

CONSIGHT-I differs from previous vision-based robot systems in its potential for practical production use. The vision subsystem, based on structured light, requires neither fluorescent belts nor high-contrast parts. Both the vision and the robot subsystems are easily reprogrammed for new parts. In fact, the simplicity of 'teaching-by-doing' is retained in spite of the complexities of the vision-controlled robot motions.

A production version of CONSIGHT-I is being built by GM Manufacturing Development. The functions handled by a single computer in the experimental system described here have been distributed among three smaller computers. The robot has been replaced with one suitable to a production environment. While the software concepts have remained the same, the implementation and programming languages have had to change to meet production requirements.

Acknowledgements

Cooperating with GM Research Laboratories throughout the project were GM Manufacturing Development and GM Central Foundry Division. Richard Elliott of GM Central Foundry Division supplied a large number of castings on which we could test our ideas and patiently described Foundry operations to us during our many visits to Saginaw and Defiance. Robert Dewar and James West of GM Manufacturing Development Staff helped design the optical system for CONSIGHT-I and loaned us lights, lenses and cameras. Most important, they offered their ideas during numerous discussions. Kenneth Stoddard designed the belt position/speed detector for the conveyor and Arvid Martin provided the system support. Joseph Olsztyn's ideas established the basis for the CONSIGHT-I vision subsystem.

References

[1] Perkins, W. A. 1977. Model-based vision system for scenes containing multiple parts. In, *Proc. 5th Int. Jt. Conf. on Artificial Intelligence*, pp. 678-684.

[2] Baird, M. L. 1977. Sequential image enhancement technique for locating automotive parts on conveyor belts. In, *Proc. 5th Int. Jt. Conf. on Artificial Intelligence*, pp. 694-695.

[3] Ward, M. R. 1976. Specifications for a computer controlled manipulator. GMR 2066, GM Research Publication, Warren, MI, USA.

[4] Agin, G. J. 1972. Representation and discrimination of curved objects. Stanford University. Artificial Intelligence Project, Memo, AIM-173.

[5] Oshima, M. and Shirai, Y. 1977. A scene description method using three-dimensional information. Progress Rep. of 3-D Object Recognition, Electrotechnical Lab., Japan.

[6] Popplestone, R. J., et al. Formation of body models and their use in robotics. University of Edinburgh, UK.

[7] Agin, G. J. 1975. An experimental vision system for industrial application In, *Proc. 5th Int. Symp, on Industrial Robots*, pp. 135.

[8] Duda, R. and Hart, P. 1973. *Pattern Classification and Scene Analysis*. Wiley-Interscience Publication, New York, p. 284.

[9] Agin, G. J. 1976. Image processing algorithms for industrial vision, Draft Report, SRI International.

[10] Paul, R. 1975. Manipulator path control. In, *1975 Int. Conf. on Cybernetics and Society*, pp. 147-152.

[11] Beecher, R. 1978. PUMA: Programmable universal machine for assembly. *GM Research Laboratories Symposium on Computer Vision and Sensor-based Robots*.

PRACTICAL ILLUMINATION CONCEPT AND TECHNIQUE FOR MACHINE VISION APPLICATIONS

H. E. Schroeder
EG&G Reticon, USA

There are several basic concepts used in lighting applications. A series of examples of illumination types are discussed as they relate to real applications. Some types of illumination techniques include: front lighting and its relationship to surface characteristics, structured lighting and its use with both matrix and linear types of photodiode type arrays, collimated light, rear illumination using both condenser optics and diffuse illumination, and retroreflectors and their use in industrial machine vision applications.

Machine vision applications all have one characteristic in common – obtaining feature information from the scene. The term used to describe this condition is 'contrast'. Contrast is created by variations in shades of grey or changes in colour. The technique used to discern these changes is a function of the item to be analysed, the optical sensor and the illumination. This paper will deal with the factors to be considered when selecting the optimum illumination technique.

Most applications using machine vision require some form of controlled lighting to obtain consistent data from an optical image. Where unstable illumination conditions exist, the signal processing grows in complexity. This material will address methods of applying various illumination techniques to enhance features and, in many cases, reduce the complexity of signal processing.

Optical reflective, absorptive and transmissive characteristics of materials

Reflective material characteristics

Diffuse. A diffuse surface is granular in nature. The incident light striking it is heavily scattered as it is reflected. Examples include non-glossy paper and textured surfaces.

Specular. This characteristic defines light striking a surface to be reflected at an angle that is equal to the incident angle. When dealing with a curved surface, the tangent must be considered at the point of impact. Examples of specular surfaces are mirrors and polished steel bearings.

Retroreflective. A retroreflective surface returns the incident light striking it back upon itself regardless of angle (within the constraints of the retroreflector design). Examples include roadway reflectors and reflective safety tape.

Spectral selective. This denotes surfaces that return uneven amounts of some wavelengths and absorb the remainder.

Spectral non-selective. This denotes surfaces that return the same wavelengths as the incident light.

Absorptive material characteristics

Non-selective spectral absorption. This condition is exhibited when all incident light wavelengths are absorbed equally. This is exhibited, for example, by a surface that is black or grey.

Selective spectral absorption. When a material exhibits a higher absorption of some wavelengths than others, the result is a higher return of some incident wavelengths than others. A coloured painted surface, for example, exhibits this condition.

Transmissive material characteristics

Transparent material. A material that transmits light radiation with no appreciable diffusion or scattering.

Translucent material. A material that transmits the majority of the incident light but has a very high diffusion component. An example of this characteristic is a viewing screen used in rear projection.

Opaque. This is a condition where a material does not transmit any light, i.e. the transmissivity equals zero.

Spectral selective. A term that defines materials that transmit specific wavelengths and either reflect or absorb the remainder. An example is coloured glass or an optical filter.

Spectral non-selective. This defines material that transmits equal percentages of all wavelengths of light incident upon it.

Spectral wavelength characteristics of illuminators

The illuminator may be defined for the purpose of this paper as the source of radiant energy that will produce a photometric-to-electrical change on the sensor.

There are many types of illumination sources. They range from sources that emit spectral energy from a single wavelength to those that produce a large envelope of wavelengths. Fig. 1 illustrates some of these sources and their respective spectral emissions.

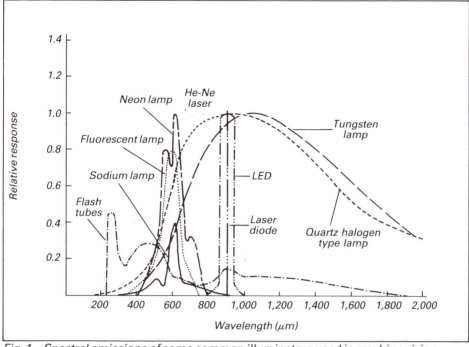

Fig. 1 Spectral emissions of some common illuminators used in machine vision

Spectral relationship of silicon-type detectors to illumination sources

The most common types of solid-state detectors used in machine vision applications are fabricated using a silicon-based device. It is important to illustrate how the spectral response of the detectors relates to the spectral emissions of the illuminators. Fig. 2 illustrates an averaged response curve for silicon. This curve will vary in shape as a function of the device structure.

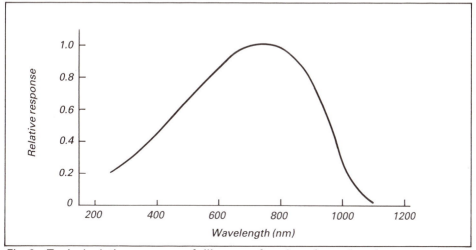

Fig. 2 Typical relative response of silicon as a function of wavelength

Self-radiators

The subject target of the sensor may well be its own light source. A subject material that emits wavelengths that the sensor will respond to is referred to as a self-radiator. An example of this may be a hot bar of steel or a stream of hot glass. These self-radiators are generally very 'rich' in the infrared portion of the spectrum and in most cases will change their emission drastically as a function of temperature. It is therefore recommended that, when feasible, the emitted wavelengths from the self-radiator should be optically blocked and a stable illumination source used. Conditions which demand that the emitted energy from the self-radiator be used can be overcome by using a real-time exposure control at the sensor. In some applications where the temperature is a factor in the process, the amplitude of the output signal from the sensor may be an aid in the control process.

Basic types of illuminators

The types of illuminators covered in this paper are grouped into the following categories. Each of these categories will be defined to clarify their use in the applications discussed, and to assist the reader in determining the specific illumination requirement. Fig. 3 illustrates the definitions.

Diffuse surface. This term refers to illuminators that have the radiant energy emitting from a translucent material. Examples of this are fluorescent lamps, light-tables and diffuse reflectors.

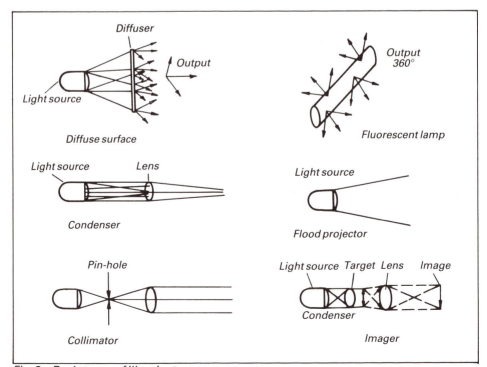

Fig. 3   Basic types of illuminators

Condenser projectors. A light source of this type changes the emitted direction of the light from an expanding cone into a condensing direction. This principle is used primarily in conjunction with imaging optics.

Flood or spot projection. This illuminator also redirects the natural expanding cone, but is used to illuminate surface areas. Examples are outdoor floodlights or car headlamps.

Collimators. In this process, the rays of energy emitted from a pin-hole are very carefully redirected to form a beam of parallel light. The beam size will not change at all in a perfect collimator. Examples of this source are lasers and optical bench collimators. Some multi-spectral collimators are commercially available for use in machine vision applications, but the amplitude of output radiation is comparatively low.

Imagers. This term refers to an illuminator that forms an image of either the lamp filament or target at the object plane. Examples of this illuminator are slide projectors and optical enlargers.

All of the above-mentioned types can be configured to operate in conjunction with matrix, linear and circular arrays. This refers to a sensor that is configured as a segmented angular ring of photodiodes[1].

Techniques

This section deals with the path of light from the illuminator to the sensor optics. Since large variations in optical imaging requirements exist, the specifics of this area are illustrated in block diagram form only.

Front illumination

The purpose of this illumination (Fig. 4) is to flood the area of interest with light such that the surface characteristics will act as the defining features in the image. This technique is used to find heavily contrasting features when working in the binary/digital field or grey-scale information using analogue-to-digital processing.

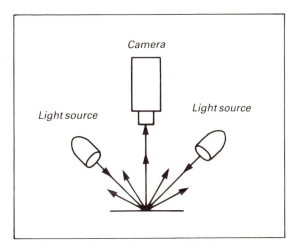

Fig. 4   Front illumination

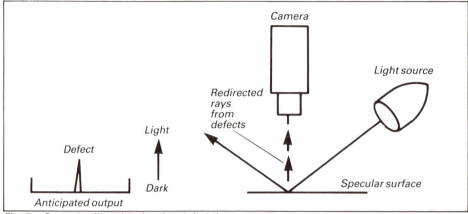

Fig. 5   *Specular illumination (dark field)*

Specular illumination (dark field)

The light striking a specular or mirror-type surface reflects off at an angle that is equal and opposite to the incident angle. The only illumination that is returned to the sensor, when it is positioned at an angle other than the reflection angle, is scattered energy from a perturbation in the surface (Fig. 5). This technique works well for surface defect recognition. The background should appear totally dark, with the defects creating the only signal amplitude change.

Specular illumination (light field)

This uses the same principle as above; however, the sensor is positioned in line with the reflected ray (Fig. 6). The only time illumination is not transmitted to the array is when a deformity of the specular surface exists. This method is also used in surface defect recognition. The uniformity of illumination is very important when using this technique, since the defect contrast may be small with respect to the entire output signal amplitude.

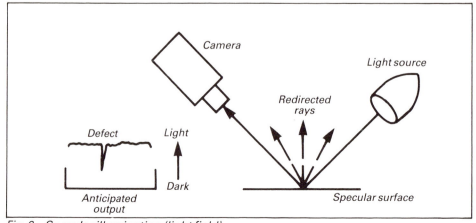

Fig. 6   *Specular illumination (light field)*

Fig. 7 Beam splitter

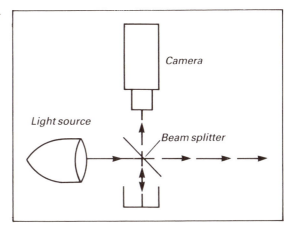

Beam splitter

A beam splitter is an optical component that has the property of reflecting some portion of the energy that strikes it and transmitting the remainder. Beam splitters appear transparent and are generally spectral non-selective. The ratio of transmission to reflectivity is established by design; however, for the sake of simplicity, we will consider a 50%/50% device. The maximum amount of source energy that can reach the sensor from the source is 25%, hence the technique is not very light efficient. The purpose of the beam splitter is to transmit illumination along the same optical axis as the sensor (Fig. 7). In this way, the sensor can view objects that would otherwise be difficult or impossible to illuminate.

Split mirror

In applications where a linear or circular sensor is to be used, a front-surface mirror with an apertured portion of coating removed will produce a result similar to the beam splitter approach (Fig. 8). This technique is far more illumination efficient, however, and hence has lower illumination intensity requirements.

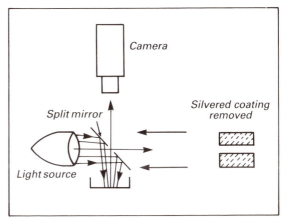

Fig. 8 Split mirror

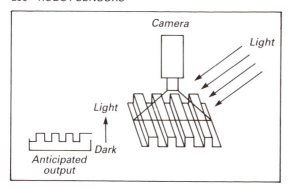

Fig. 9   Offset illumination (shadow)

Offset illumination

The light source, when offset (Fig. 9), produces shadows of features which define the location of the feature. This technique may be used in raised bar code applications or a hole sensing task.

Non-selective redirector

This technique is used when the source does not produce the proper illumination geometry and must, therefore, be redirected. Two basic examples of this are the 'secondary diffuser' and the 'broken mirror'. These two examples are explained in the section on applications below.

Rear illumination (lighted field)

With this technique, the primary type of illuminator used is a surface diffuser and the sensor is located on the opposite side of the subject (Fig. 10). Uses of this technique include silhouette feature inspection of parts and web width measurements. (Note: In almost all rear illumination applications, a silhouette of the subject is produced by simple obstruction. When grey scale is used, either the subject will have a specular surface or the subject will be transparent or translucent.)

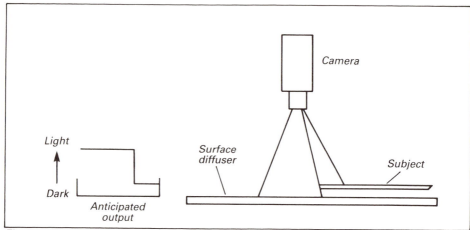

Fig. 10   Rear illumination (lighted field)

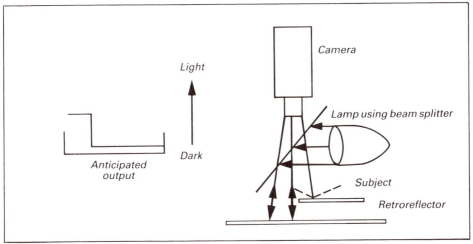

Fig. 11   Retroreflector

Retroreflector

A retroreflector is an optical device that redirects the incident rays back in the general direction they came from. The incident angle can be varied considerably depending on the design of the retroreflector. This technique allows the light source to be on the same side of the subject as the sensor (Fig. 11). Several variations in the technique are possible, such as using the specular reflection characteristics of a subject to enhance the contrast (specular illumination dark field; Fig. 12).

Double density

The resultant energy of light passing through a medium is equal to the incident energy times the transmissivity of the medium. If the energy must pass through the medium twice, as illustrated in retroreflector techniques, the energy at the sensor will be the incident energy times the transmissivity squared.

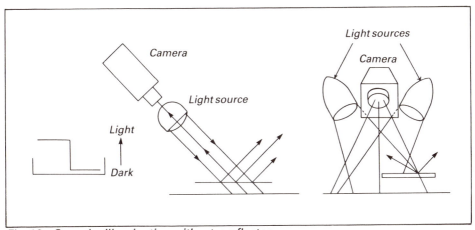

Fig. 12   Specular illumination with retroreflector

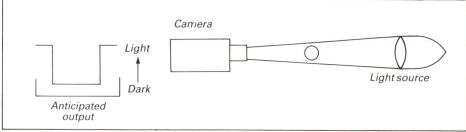

Fig. 13   *Rear illumination (condenser)*

Rear illumination (condenser)

A condenser, when used for rear illumination (Fig. 13), must form a condensing cone that is larger than the sum of the subject size and the required clearance for part movement. This technique may be used in conjunction with a specular surface and will maintain the same characteristics and limitations. A condenser system as a back light is very efficient and will produce a higher-contrast image than a lighted field. When possible, this technique should be used for high-magnification applications.

Rear illumination (collimator)

Collimated light can be used both with and without an imaging system. Since the source produces parallel light rays, and the effect of the obstructed rays is a silhouette, the location of the object on the axis between the source and the sensor is flexible (Fig. 14). This technique is used in applications where all edges of the object are not in the same object plane or the subject may vary in its object plane location.

Front illumination (imager)

This technique is used for applications where position or depth information is required. 'Structured light' applications fall into this category (see Fig. 15). The purpose of this technique is to create a target of imaged light superimposed upon the subject surface. The pattern this light forms on the target, as viewed by the sensor, contains the information required to satisfy the application requirements. An example would be a beam intersection displaced as a function of the subject thickness. This process is called triangulation.

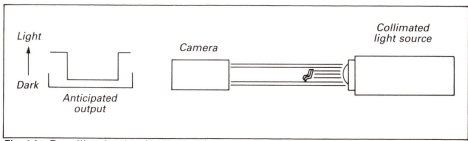

Fig. 14   *Rear illumination (collimator)*

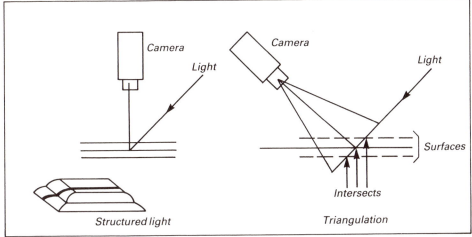

Fig. 15   Front illumination (imager)

Rear offset illumination

When the feature of interest is located in a transparent medium and the feature will scatter the light rays, this method can be used to produce feature highlights similar to those of dark-field specular illumination. The technique is illustrated in Fig. 16. The advantage over direct rear illumination is the contrasting characteristic of the light feature on a dark background as opposed to a grey feature on a light background. This technique can also be used with translucent materials having opaque features but is not as effective. The output from the sensor when used on a translucent medium will be the opposite to that on a transparent medium.

Design considerations

When considering the appropriate illumination for any application, several factors must be addressed. This section will cover some of these and include questions to help the reader to implement applications successfully.

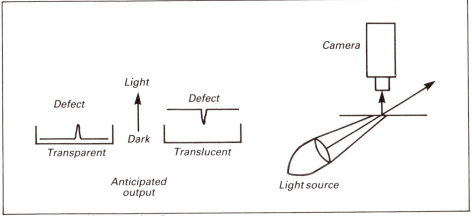

Fig. 16   Rear offset illumination

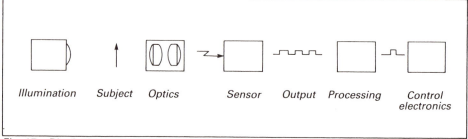

Fig. 17   Block diagram of system

System interaction

When considering the use of machine vision, the total system should be addressed. A block diagram of the parts should be made (Fig. 17) and each function should be examined to verify its effect on all other functions.

The interaction of the illumination source with the remainder of the system will be briefly discussed here. Each category below is comprised of questions that should be answered by the machine vision system designer.

Illumination to subject:

- What information is required from the feature?
- What are the specular, absorptive and diffusive characteristics of the feature/background?
- How large, or small, is the subject?
- What are the spectral characteristics of the subject?
- What is the geometry of the subject and how does the feature fit in that geometry?

Illumination to optics:

- Does the illumination's geometry properly fulfil the lens requirement?
- Is the specular output of the illumination consistent with the optical capabilities?
- Are the optics capable of transmitting the required illumination intensity to the sensor?
- Do the optics reproduce the feature information per the illumination technique's algorithm?

Illumination to sensor:

- Is the illumination of sufficient intensity to produce the desired response?
- Is the illumination technique adequate to form the desired image on the sensor?
- Is the spectral output of the illumination source compatible with the sensor's spectral response?
- Is sufficient contrast created by the illuminator to be defined by the sensor?
- Is uniform illumination obtained on the sensor where applicable?
- Is the illuminator capable of providing the necessary spectral output where colour separation is required?

Illumination to output:

- Is the output of sufficient amplitude to produce a usable sensor output?
- Is the illumination source stable enough to provide a consistent output?

Illumination to processing:

- Has the proper illumination technique been selected to produce the required information?
- Is the illumination stable enough to provide consistent results?
- Is an illumination feedback or exposure control required?
- Are false data produced by unwanted secondary illumination sources or specular reflections?
- Has synchronisation been taken into account when fluorescent, pulsed or timed illumination is used?

Illumination to control electronics:

- Is the illumination source obstructed or in any way affected by any control mechanisms?
- Must the control electronics employ any other inputs to create pulses or timed illumination?
- Is a fail indication employed in the event that a lamp burns out, degrades in intensity or otherwise fails?

Reliability

The source should be selected to produce the desired output under the constraints of the application. Some of the factors that affect this are:

- The mechanical stability of filament lamps in vibrational environments.
- The decay in intensity or change in colour temperature as a function of ageing.
- The mechanical stability of the fixturing used to implement the technique.
- The changes in light output caused by foreign material, i.e. oil, dirt, fumes, etc.
- Variation in light output when lamps/sources are changed.

Applications

The following application examples illustrate how an understanding of the application and the constraints placed on it can help the designer to pick the best technique.

Surface inspection of bearing rolls

A bearing roll has a specular surface. The geometry of the roll is similar to that of a barrel. The defects to be sensed include nicks, cracks, scuffs, pits, scratches and flats. The current inspection is done visually by humans using a 5 × magnifier and rotating the roll about its optical axis while looking for distortions in the light reflected off the roll.

The first consideration made when selecting the illumination was the change in specular characteristics caused by the defects. The second was the

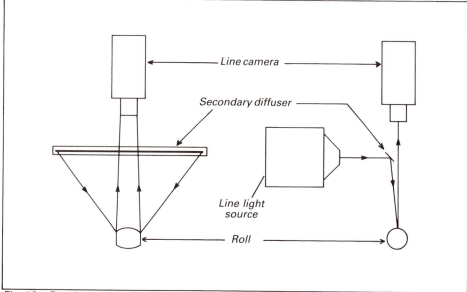

Fig. 18   Bearing roll defect inspection

barrel shape, since illumination from a condenser or a source with a projected beam would not provide even illumination across the entire length of the roll.

The technique chosen was specular illumination from a lighted field. The sensor used was a linear array camera imaging the edge of the reflected secondary diffuser (Fig. 18). The roll was then rotated on its axis to 'inspect' the entire roll surface. The ends of the roll were also inspected using specular illumination, but because the surfaces were flat, a condenser illuminator was used as the source. This system is in factory use with excellent results.

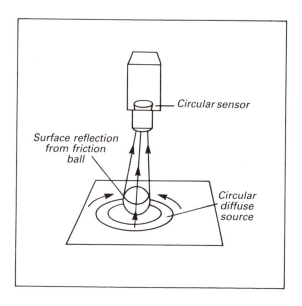

Fig. 19   Spherical friction ball inspection

Friction ball surface defect inspection

This application is similar to the roller bearing problem except that a ball or sphere has a far greater radius of curvature. In this situation, the variable optical signatures from the defects at the different object distances to the sensor required the use of a circular scanning technique. An apertured lighted field using specular illumination was chosen. In this way, the sensor's elements are all equidistant from the inspected surface. The result is a consistent defect signature. When the friction ball is rotated in an elliptical fashion, the entire surface can be inspected in a minimum period of time (Fig. 19).

Broken mirror illumination

This application required the subject illumination to be scattered such that even multi-directional rays would strike it. To accomplish this, a diffuse source would work; however, not enough illumination could be obtained. To produce the required illumination, a reflector was formed of many pieces of broken mirror epoxied in place at irregular plane angles. When a simple projection lamp was directed at the reflector, the light was scattered in all directions onto the subject surface. This technique gave both adequate scattering and adequate intensity.

Carton sizing using offset optics and retroreflection

The following constraints made this application interesting:

- The cartons would vary in height, width and length from carton to carton.
- The location and configuration of the conveyor made mounting a light source under the cartons impractical.
- The cartons could be any colour.

It was possible to register the cartons against one surface other than the base. An update sensor could be installed to indicate when the edge of the carton passed. Minor modification to the conveyor roller mechanism was allowed, but had to be easily reproducible on similar existing conveyors.

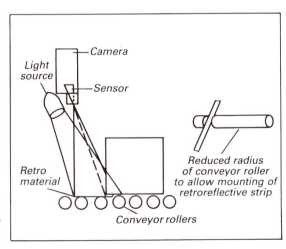

Fig. 20   Carton sizing using offset optics and retroreflection

The problem of distance variations caused by carton size variations was solved by imaging the bottom edge of the carton using retroreflective illumination (Fig. 20). The retroreflector was made of tape and mounted on a thin strip. The strip was mounted between the conveyor and the carton by cutting a section of the rolls to a smaller radius. The optics were offset to put the optical axis in line with the end of a line sensor. A wide aperture sensor was used to 'average' the illumination along that axis and minimise the effects of dust and dirt. The distance from the sensor field and the update sensor was calibrated and a fixed offset added into the measurement calculation. (Note: Very high contrast is obtained by using a retroreflector behind a white sheet of paper.)

Acknowledgements

I would like to express my gratitude to all of the companies that have shared their machine vision problems with me. I would also like to thank EG&G Reticon for giving me freedom to explore the feasibility of machine vision applications for the last ten years.

References

[1]   Schroeder, H.E. 1984. Circular scanning and its relationship to machine vision, In, *3rd Annual Machine Vision Conf.* SME, Dearborn, MI, USA.

THREE-DIMENSIONAL LOCATING OF INDUSTRIAL PARTS

R.C. Bolles
SRI International, USA

Locating a particular part among a pile of other parts and presenting it, correctly positioned and orientated, for processing is a difficult process to automate because of the complexity involved. The objective of the program described is to develop general-purpose techniques to locate partially visible objects, the first step towards full automation of the 'pick-and-place' operation. In the experimental system under development, a structured-light range-finder and data processing techniques are used to identify parts on the basis of their characteristic features.

One of the factors inhibiting the application of industrial automation is the inability to acquire a part from storage and present it to a workstation in a known position and orientation. A classic example of this is the 'bin problem'; workers have to reach into a bin in which the stored parts are all jumbled together (Fig. 1), retrieving one part at a time as needed. Automating this acquisition process is difficult because a part in a bin can appear completely different to a sensor, depending on its position and orientation, and some of its features can be occluded by other parts, depending on its relative position within the stack.

Our programme goal is to develop general-purpose techniques for locating partially visible parts with three-dimensional uncertainties in their positions and orientations. Our approach is to use three-dimensional part models to interpret range and intensity information. Our rationale is that, first of all, range data simplify the locational analysis because the geometric information is directly encoded in the data. Secondly, range sensors will soon be economical for industrial tasks. Finally, familiarity with the model of a part will add enough new constraints to make it practical to locate relatively complex parts jumbled together in a bin.

Since only a subset of a part's features will be visible, the basic approach of the locational system will be to acquire data, locate features in the data,

Fig. 1   A bin of castings

match clusters of features, use the information derivable from the clusters of features to gather more data and locate more features, and so on.

Our research strategy is to:

● Design a set of models to represent important features of three-dimensional industrial parts, such as cylindrical holes and protrusions.
● Develop mathematical techniques to locate instances of these models in range and intensity data.
● Investigate techniques to locate clusters of features.
● Develop a straightforward technique to acquire range data.
● Implement computer programs that utilise models to locate a large class of workpieces.

Research results

This project began in March 1980. Our first concern was to obtain high-resolution, low-noise 'range pictures' from real parts such as castings. Although such range sensors are not currently available, we believe that they will be in about five years. Therefore, we decided to implement a simple

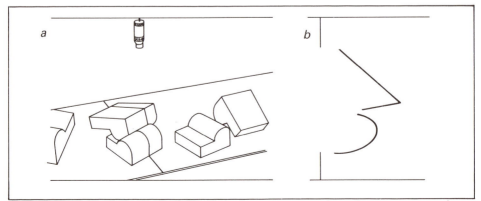

Fig. 2   A simple range sensor: (a) design of a range sensor based on a camera and a plane of light, and (b) image of the intersection of the light plane and the objects

range-finder based on structured light that is similar to the ones developed at the University of Edinburgh and other places. Fig. 2 illustrates the basic concept. The objects are incrementally moved underneath a calibrated camera and light plane and triangulation is used to compute the (x,y,z) coordinates of the points along the intersection of the light plane and the objects. A more detailed description of this device and its calibration is given in the following subsections.

Our second concern was to identify some of the most common features of industrial parts and to develop mathematical models that could be used to locate occurrences of them in range data. The list of common features includes planar regions, dihedral angles, cylindrical holes and cylindrical protrusions. Since several techniques have been developed to locate planar patches, we decided to concentrate initially on cylindrical features. Our first demonstration will be to locate the cylindrical portions of castings in a bin. This demonstration has not been completed. However, some of the mathematical models and preliminary matching results are described in a later subsection.

The range-finder

Fig. 3 shows the hardware set-up for the range-finder. A General Electric TN2500 camera, a solid-state camera with a spatial resolution of 240 by 240 pixels, looks down at objects on an XY table. A cylindrical lens spreads the beam from a 15mW helium-neon laser into a plane that intersects the objects in front of the camera. The light source is not visible, but the intersection of the light plane and the black backdrop shows its relative position.

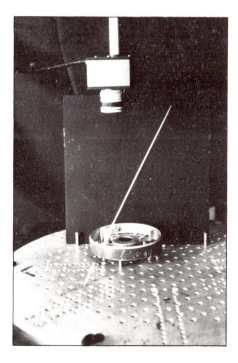

Fig. 3   The range sensor

The low-level processing of an image of the intersection of the light plane and the objects is performed in several steps, some of which are optional:

- Threshold the grey-scale image to form a binary image.
- Group together pixels that are 'on' to form connected regions.
- Thin the regions to form broken line segments.
- Filter out noise and secondary reflections.
- Classify the segments as lines or curves.
- Compute the (x,y,z) positions for points on the intersection.

Fig. 4 shows examples of some of these steps. The threshold used in the first step is generally set low in order to include the reflections off the sides of curved objects. However, as a consequence, some secondary reflections, such as the rightmost region in Fig. 4(a), are bright enough to exceed the threshold. To delete some of these reflections we have implemented an optional filter. Since the objects are opaque, each ray of light can only intersect them once. Therefore, if two segments in an image overlap in a way that implies that a ray intersects the objects twice, the two segments are marked as incompatible with each other. When an image contains incompatible segments, the filter constructs the list of mutually compatible segments that corresponds to the longest intersection. The assumption behind this heuristic is that the longest (and possibly strongest) intersection is the actual intersection. Notice that the rightmost region in Fig. 4(a) was deleted because the alternative was longer and was compatible with the same set of segments.

We have implemented a semi-automatic procedure to compute the information required to perform the filtering described above and compute the (x,y,z) positions of points along the intersection. Three pieces of information are necessary: the camera matrix that maps (x,y,z) 'world' coordinates onto the image plane, the equation of the plane of light, and the position of the light source in the plane of light. The camera matrix and the equation of the plane are combined to form a 4×3 homogeneous matrix that maps points in the image plane that are on the intersection onto their

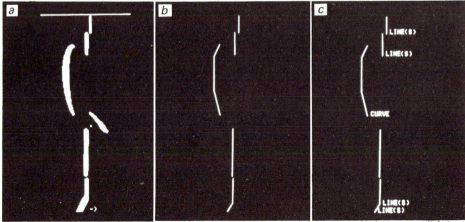

Fig. 4   *The results of three low-level processing steps: (a) the threshold image, (b) the filtered data, (c) the classified segments*

corresponding (x, y, z) coordinates. Therefore, given the image coordinates (u, v) of a point on the intersection, the corresponding (x, y, z) coordinates are computed as follows:

$$\begin{pmatrix} x^1 \\ y^1 \\ z^1 \\ s \end{pmatrix} = \begin{pmatrix} m11 & m12 & m13 \\ m21 & m22 & m23 \\ m31 & m32 & m33 \\ m41 & m42 & m43 \end{pmatrix} * \begin{pmatrix} u \\ v \\ 1 \end{pmatrix}$$

$$x = \frac{x^1}{s}, \qquad y = \frac{y^1}{s}, \qquad z = \frac{z^1}{s}$$

Since this computation involves only eight multiplications, eight additions and three divisions, it can be performed quite rapidly.

Fig. 5(a) shows a typical range picture taken by the range-finder. It is a composite of 42 slices at the end of a two-inch cylindrical casting taken a tenth of an inch apart. Fig. 5(b) shows the same data as in Fig. 5(a), but rotated so that the viewing axis is approximately along the axis of the cylinder.

An experiment to locate cylinders with a known diameter

The goal of this experiment is to locate cylinders in a jumble, given their diameter and range data taken from above them. Previous research has produced ways to identify and locate cylinders in range data. The key question for this research is how to capitalise on the fact that the diameter is known in advance. For example, is there any way to convert the problem into a one- or two-dimensional clustering problem that can be performed quickly?

The intersection of a plane and a cylinder is an ellipse in the plane with the following properties:

- Its centre lies on the axis of the cylinder.
- Its minor diameter is equal to the diameter of the cylinder.

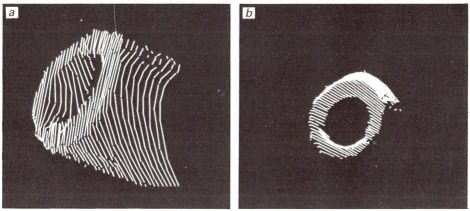

Fig. 5 A picture constructed from 42 separate slices taken at the end of a cylindrical casting: (a) the filtered data as seen by the camera, (b) the same data as in (a), but viewed from a point on the axis of the cylinder

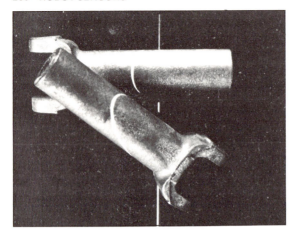

*Fig. 6   The intersection of the
light plane and two
cylindrical castings*

- The ratio of its major diameter to its minor diameter is the cosine of the angle of inclination of the axis of the cylinder with respect to the plane.
- Its orientation in the plane in conjunction with its centre and the implied angle of inclination completely determine the position and orientation of the cylinder with respect to the plane.
- The ellipses formed by two parallel planes have the same shape and orientation, but are shifted relative to one another by an amount that depends on the distance between the two planes and the orientation of the ellipses.

Our first program to locate cylinders uses this information as follows: it fits ellipses with a fixed minor diameter to all curve segments as they are located by the range-finder, and then it locates peaks in a histogram of their orientations and/or the sizes of their major axes to hypothesise possible occurrences of the cylinder.

This method reduces the problem to a one- or two-dimensional histogramming problem, but it is difficult to fit ellipses to the curves in the light plane. Since the data are taken from above, only the top half, at most,

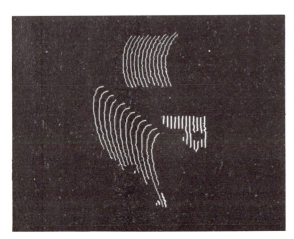

*Fig. 7   Twelve slices along the
two castings shown in
Fig. 6*

Fig. 8 Circles fitted to two curve
segments as seen from
the camera's viewpoint

of each cylinder is visible. In fact, because the camera and light source do not 'see' exactly the same parts of the scene and because the apparent width of the intersection line decreases as it curves down the side of a cylinder, only about one-third of the actual ellipse is detectable from an unoccluded cylinder. At present we have implemented three different ways to fit ellipses to such data and are currently investigating their strengths and weaknesses.

Since it is easier to fit circles with a known diameter than it is to fit ellipses, we try to fit a circle to each curve segment before trying to fit an ellipse. If the circle fits the data well, it implies that the cylinder is almost perpendicular to the light plane. In that case, the centre of the circle is put in a list for later processing. If the circle doesn't fit well, an ellipse is tried. If the ellipse fits well, a pointer to it is placed in a bin of a histogram. If the ellipse doesn't fit, the curve segment is discarded. After processing all the curve segments, the histogram and list of circles are examined for clusters large enough to suggest possible cylinders.

This demonstration has not quite been completed. The ellipse fitting needs to be improved, and the histogram analysis has to be added. Figs. 6 through 10 illustrate the current status of the program. Fig. 6 shows the

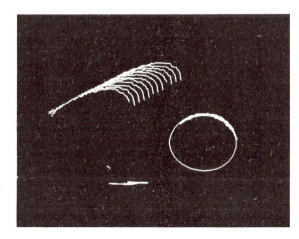

Fig. 9 Circles fitted to two curve
segments as seen from a
point on the axis of the
cylinder

Fig. 10   A circle and two ellipses fitted to three of the elongated curve segments

intersection of the plane of light with two cylindrical castings, one of which is leaning on the other. Fig. 7 shows a sequence of 12 slices of these two castings as seen from the viewpoint of the camera. The group of symmetric intersections at the top is from the casting lying on the table; the group of elongated ellipses is from the other casting; and the small group of line segments is from the top of the *XY* table that has holes in it. The symmetric intersections are almost circular in the plane of light and the circle-fitting step fits a circle to each one of them. Two of these circles are shown in Fig. 8 (as would be seen from the camera). They are shown again in Fig. 9, which presents the data as seen from a camera looking along the axis of that cylinder.

Fig. 10 shows three of the curves fitted to the segments from the 'top' casting. The segment at the bottom right is approximated by a circle because the elongated ellipse happened to be broken up into two pieces, one of which was too small to be considered, and the other could be roughly approximated by a circle. Later processing would have to filter out such mistakes. The other two ellipses, which are typical of the other cross-sections, were obtained by fitting general conics to the data and checking to make sure that they were ellipses and that their minor diameters were relatively close to the expected value. Future versions of this program will employ algorithms to fit ellipses that have known minor diameters.

This demonstration, although incomplete, is an example of the basic theme of our research: use models of the objects to quickly and reliably locate occurrences of them in range data.

Program objectives for the future

Our immediate objective is to complete the program to locate cylinders. To do this we are investigating ways to fit ellipses that have known minor axes to data that only cover from a quarter to a third of the ellipse. We are also experimenting with better region-thinning algorithms that the selection of the midpoints of each horizontal segment, which we currently use. Midpoints are fine for sections of the intersection curves that are essentially vertical in the image, but they lead to inaccurate and sparse data for sections that are almost horizontal.

Our next objective is to develop models for the different types of range discontinuities, such as occlusion edges, tangential edges and intersection edges, and use them to locate occurrences of specific object features such as the ends of protrusions and dihedral angles. The next step is to investigate structure-matching techniques to locate clusters of object features that can be used to hypothesise occurrences of objects. And finally, we plan to consider verification techniques to test these hypotheses.

VISUAL GUIDANCE TECHNIQUES FOR ROBOT ARC WELDING

C. G. Morgan, J. S. E. Bromley, P. G. Davey and A. R. Vidler
Meta Machines Ltd, UK
(formerly of the University of Oxford)

The metal-inert-gas (MIG) arc welding process is critically dependent upon the accurate positioning of the heat source (arc) and upon maintaining optimal welding parameters (travel speed, wire-feed rate, voltage, etc.). These constraints have until now restricted the range of application of conventional sensorless arc welding robots. Recent work at Oxford has produced a visually guided system, operating in a single pass, which corrects for typical variations in seam centre line and gap which occur in the arc welding of thin sheet steel pressings. This system uses a structured-light technique and is unique in that the complete sensor, consisting of two CCD area cameras and four laser diode light sources, is packed around the welding torch in a cylinder only 57mm dia. × 200mm long. The complete sensor-guided arc welding robot was demonstrated to industry for the first time in March 1983. This paper describes the hardware and software design of the real-time vision processing system, including ruggedisation of the sensor in order for it to survive close to the arc, the design of a hardware 'stripe finder' for preprocessing the structured-light image, and progress in generalising the software algorithms to cope with the full range of joint situations encountered, under conditions of strong visual interference.

The industrial use of robots for arc welding is at present limited because of variations in the fit-up of components. There are many sources for this lack of uniformity including wear on jigging and press tools, springback after pressing, age hardening and thermal distortion during the actual welding. There are two possible solutions to this problem. The first is to improve the quality of the components, which would entail better jigging and extra pressing operations together with improved stacking, transportation, inspection, maintenance and quality control. The alternative is to use a sensor-controlled welding system which is able to adjust automatically to variations in the actual seam positions and fit-up conditions. An earlier study[1] discussed the relative merits of each approach and showed a large economic advantage in favour of the sensor-based solution if it could be achieved. Our progress since then is described in a subsequent report[2]. Work on the sensor-guided arc welding problem has also been reported from a number of other sources listed in the references [7-9].

The work at Oxford has taken as its focus the arc welding problems typically encountered in the motor car industry. Components consist of thin sheet steel pressings (1-2mm thick) of complex 3-D shapes with no joint preparation. The welding must be repeated on large numbers of components, and fit-up uncertainties of the order of 3mm need to be corrected to better than 0.5mm for a good weld. In its present form the system is not used to track unknown seam geometries; rather, it continually makes small corrections to an expected or pre-taught path. To be economic the welding must be completed in the minimum possible cycle time, implying a single-pass solution with inflight adaptive control. Both the jigging and the nature of the components make it difficult to obtain access for welding; this can only be achieved by a compact sensor unit. Typical welding speeds are in the range 5-20mm/sec. This places a requirement on any closed-loop system of being able to sample the seam position at a rate of at least five times per second.

A number of design problems must be solved if a viable system is to function in the severe environmental conditions posed by the welding process. These include:

- Arc glare (electrical welding currents of hundreds of amps create an intense plasma of rapidly fluctuating intensity).
- Weld spatter (blobs of molten metal fly across the field of view).
- Smoke and fumes.
- Heat.
- Electrical noise.

The vision problem; structured light

The sensor operates on the principle of active triangulation ranging, a technique which many researchers have used to obtain information about the three-dimensional layout of surfaces in the scene[3-5]. A sheet of light is

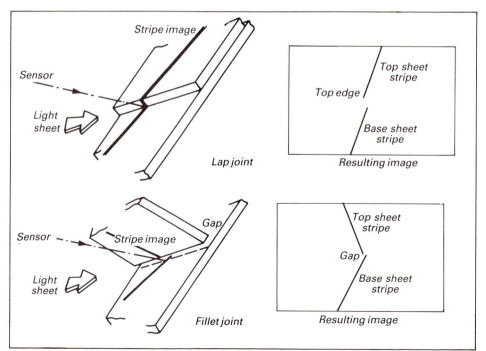

Fig. 1 *Stripes from common joints*

generated using a combination of a laser diode and a cylindrical lens. This sheet of light is arranged to fall across the camera field of view at a known angle. The image resulting from the intersection of this light sheet with the workpiece is of two curved or straight-line stripe segments corresponding to the two pieces of metal to be joined. The sensor is located immediately in front of the welding torch. Fig. 1 shows the nature of the images formed from the two most common weld joints encountered, the lap and the fillet or T-joint. Each stripe position in the image maps to a corresponding 3-D position on the workpiece. The position of the metal sheets, and hence the position of the joint, can be found by analysing the stripe sections. The tracking error can be determined by comparing this measured joint position with some taught position. Both lateral (side-to-side) errors and stand-off (distance between welding torch and workpiece) can be measured. In addition to these positional errors, the stripe analysis also allows the seam gap or fit-up error to be determined.

The problem of observing the image resulting from the light stripe is complicated by a number of factors, most particularly the high level of spurious information in the image. The most commonly observed spuriae may be characterised as follows:

- *Arc glare:* The light radiated from the arc and the molten weld pool beneath it is very intense. The emission is also very broad band, with the peak energy density in the UV. The wavelength sensitivity of the camera is very broad, extending from the near UV to beyond 1000nm in the IR. The laser output is, however, virtually monochromatic at a wavelength in the region of 830nm in the near IR. This allows the use of a very narrow band interference filter (whose half-power bandwidth is approximately 8nm) tuned to the laser's wavelength in order to reduce the intensity of the unwanted background illumination by at least two orders of magnitude. Further reductions, which are described in a following section, are made by the mechanical design of the sensor. Nevertheless, the arc glare is not completely removed and it may still be brighter than the stripe in parts of the image.

- *Weld spatter:* As a side-effect of the MIG process, small beads of molten metal (typically 1-2mm dia.) are ejected from the weld pool. Some of these particles will of course fly across the field of view of the camera. Since they are red hot they radiate a significant amount of energy in the near IR and are therefore visible to the camera/filter combination. Traces of such particles appear on a majority of the images we have examined, and in all cases they were travelling fast enough to have crossed the field of view completely in one frame period of 20ms. As they travel above the object plane they produce out-of-focus images. In the CCD imaging device we use, the effect of the fast passage of a luminous object in this manner is the addition of a 'trail' of increased brightness to the original image. Since the camera field of view lies ahead of the weld pool along the seam direction, and the spatter emanates approximately radially from the weld pool, these trails are (with very few exceptions, caused by the spatter bouncing off obstructions) approximately parallel to the seam direction, in contrast to the stripe which is within 25° of perpendicular to the seam direction. In addition, the trails tend to be much broader than the stripe, and this aids discrimination between the two.

- *Specular reflection:* The biggest problem results from the nature of the metal surfaces, which act more as specular reflectors than as diffuse scatterers. The resulting image varies considerably in intensity depending upon the angles of the joint. The rather common T-joint configuration can act as a corner reflector, giving rise to a strong reflection of the stripe from each metal surface on the other. This spurious reflected stripe is often more intense than the original projected stripe, but it is badly out of focus and therefore much wider than the original.

These facts make the use of simple image thresholding techniques ineffective. However, it is evident that all of the above spuriae can be rejected by the use of a spatial filter. Further, such a filter need only operate in the direction parallel to the seam, since the wanted stripe is known to lie roughly perpendicular to it and thus has a well-known brightness profile as measured parallel to the seam. In early software trials of such spatial filters, a number of non-linear filtering methods were evaluated, but the use of the one-dimensional linear difference-of-Gaussians (DOG) filter has proved quite satisfactory in practice and has the advantage that, being a linear operation, it can readily be implemented in hardware. An inexpensive device, described below, has been developed to carry out the filtering, operating in real-time along each video raster line.

Development system overview

The major components of the development system are shown in Fig. 2. The welding torch/sensor package is carried by a conventional industrial robot. At present we use an ASEA IRb-6 robot, which has five degrees of freedom and a payload capacity of 6kg. The robot is controlled by the standard ASEA controller and computer interface to which a number of minor modifications have been made to permit direct control of the axes by an external computer. A sixth axis has been added which allows the sensor unit to be rotated about the torch axis. In this way the sensor can always be orientated along the direction of the weld seam. The computer interface has been extended to give the adaptive control performance necessary. The welding power is supplied by an Aga MIG-welding set modified to allow computer control.

The processing tasks are divided between two DEC PDP-11/23 computers, one being dedicated to robot control and programming functions and the other to vision processing. As the early versions of the ASEA controller did not have a coordinate conversion capability, we have implemented a kinematic model and control subsystem to permit the programming and execution of robot programs in terms of Cartesian world coordinates.

The vision computer is fitted with a frame store which can digitise and store the video signal as a 256-256 array of pixels, with eight bits of grey-level information per pixel, and a hardware stripe finder to convolve the video

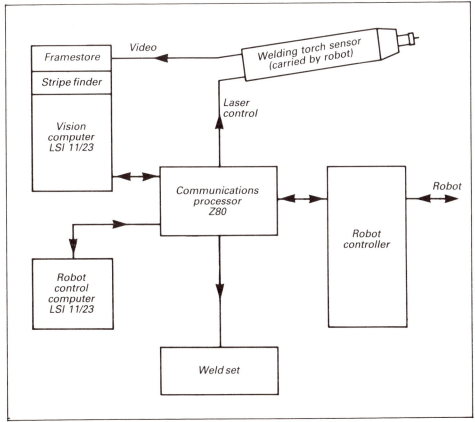

Fig. 2   Block diagram of equipment

signal with a DOG filter in real-time. The intensities of the laser light sources can be individually controlled for optimal image quality; this enables us to cope with the wide range of stripe intensities resulting from the different joint situations.

The communications subsystem is based on a Z80 microprocessor running a real-time multi-tasking executive that is responsible for coordinating the computations performed by the control subsystem, vision subsystem and robot controller.

The sensor

Fig. 3 shows a cross-section of the sensor unit with CCD camera and stripe generators mounted around the welding torch in a cylinder 57mm in diameter. The camera is based on a 488×380 pixel CCD area array and yields CCIR standard 625 line video output. The field of view on the joint is approximately 16×19mm centred 16.5mm in front of the arc. The light source is a laser diode and cylindrical lens combination mounted on each side of the torch, generating a light sheet which intercepts the camera axis at an angle of 18°. The shielding gas shroud surrounding the weld tip serves the secondary function of creating a shadow zone which prevents the direct

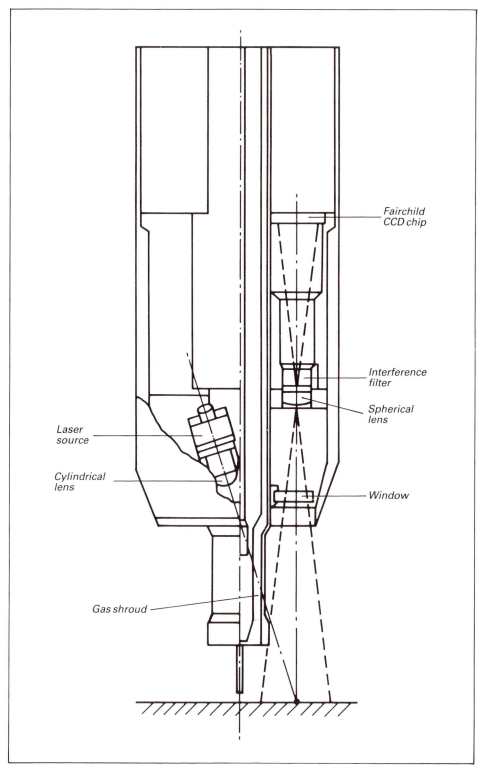

Fairchild
CCD chip

Interference
filter

Spherical
lens

Laser
source

Cylindrical
lens

Window

Gas shroud

Fig. 3   Torch/sensor cross-section

arc-glare from entering the optics and which also greatly reduces the bombardment of the camera's protective window by the flying weld spatter. This window covers the sensor openings and, together with an air bleed, protects the optics from fumes, spatter and smoke. When the window becomes dirty the robot can rotate it to expose a clean area until it is finally replaced. Trials in a production environment have indicated that the life of this window before its eventual replacement will be in excess of eight hours' continuous welding, which is comparable to the life of the standard 14kg reel of consumable welding filler wire. A twin water cooling system extends the length of the sensor to maintain the operating temperature at 20K or less above ambient. Electronics necessary to control the camera chip and lasers are mounted in the space above the camera.

The design of the sensor allows for a second identical camera/striper arrangement to be mounted on the opposite side of the torch. This enables observation of the cooling weld pool immediately after welding to provide a means of rapid quality control. Research on these techniques will extend as the project continues.

The stripe finder

The hardware stripe finder has been implemented using a commercially available tapped analogue delay line integrated circuit as the basis of a transversal filter. The conversion from the spatial to the temporal domain is of course effected by the camera line scan and this obliges us to arrange the camera so that its line scan is parallel to the seam direction.

The dedicated image preprocessor used to perform the filtering and allied operations has become known as the 'hit box'. Its purpose is to operate on the raw image data (which in the present implementation are represented as a 256×256 pixel array of 8-bit brightness values) and reduce it to a much smaller array having only 256 elements, one for each line of the video image, referred to as the hit list. Each element of the hit list contains one 8-bit value identifying the picture column in which the strongest response to the DOG filter was found on that line, and one value (currently of 6-bit resolution) containing the strength of that response. Identification of the column in which maximum response occurs is achieved by loading the column number (from a binary counter) into a latch whenever the response exceeds the previous maximum which is stored in a sample-and-hold. At the end of each video line, the column and strength values are strobed into a further latch from which they may be read into the computer at any time during the next $64\mu s$. By this means the filtering and peak-detection operation is completed in one frame period of 20ms. The transversal filter directly performs the convolution of the input signal with its impulse response, which can be given an arbitrary shape by adjusting the potentiometers which control the weighting coefficient at each tap of the delay line.

Processing stages

The operation of the system can be divided into four basic stages, as shown in Fig. 4 and as discussed below.

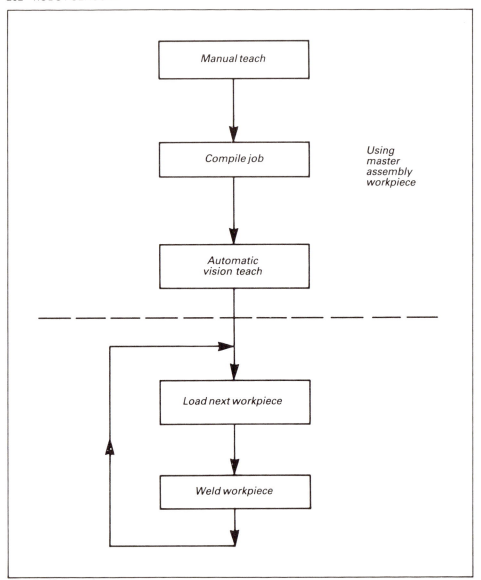

Fig. 4 Processing stages

Manual teach

A pre-production master assembly is loaded into the jig and the job is taught. The robot is manipulated by means of a standard teach pendant to position the welding torch at the start, end and optional intermediate points along the weld seams. Commands are issued to the control computer to record these positions, together with information defining the name, configuration, sense (left or right handed) and metal thickness of each joint. Robot approach points and speeds are also recorded. Pathways connecting seams and approach points are defined and a short program written defining the required welding sequence.

Job compilation

The job program is automatically checked for consistency and to verify that all robot movements can be made along known safe pathways. Intermediate positions along seams are precomputed together with Jacobian matrices relating observations in the camera frame of reference directly to joint angle coordinates. These calculations are designed to minimise the work-load to be performed at weld time.

Automatic vision teach

The robot automatically makes a slow traversal along each weld seam in turn. Video frames are collected and analysed at intervals and joint description records defining joint location and stripe angles are constructed. Checks are made for the conformity of the image to the joint type and for spatial continuity.

Adaptive welding

The program to weld the job is now executed on successive real workpieces. The control computer informs the vision processor which seams to weld and moves the robot to the start-of-seam position. The vision processor enters a 'pre-weld' mode in which video frames are collected and processed. Joint positions are compared with the teach time situation and positional errors and gap information transmitted back to the controller. Robot position corrections are made until the controller is satisfied that the torch is correctly located at the start of the seam. The welding sequence is then initiated and the vision processor enters the 'weld' mode. Video frames are captured and processed by the stripe finder. Using information supplied by the communication processor each video frame is identified by a value representing the distance traversed by the torch along the weld seam. The corresponding teach time joint record is located in the database. The image data is analysed and positional errors between the teach time model and the actual weld time image, together with the joint gap measurement, are fed back to the control computer in order to close the feedback loop. The control program utilises its knowledge of the joint type, metal thickness and gap size to select appropriate welding parameters using a model based on work carried out by the Welding Institute[6]. Robot trajectories are corrected using an appropriate feedback algorithm. This process is repeated until the end of the seam is reached. Currently, the system allows about five frames per second to be processed.

Vision processing

There are two vision processing stages to consider: the teach phase and the welding phase. The two stages are quite distinct and require different processing approaches.

Vision teach processing

The vision teach phase is not time critical, and no welding is in progress. The workpiece being examined is a master assembly which can, if necessary, be painted matt white in order to give the best possible image quality. Video frames are captured together with their corresponding hit list. Processing is as far as possible confined to the use of hit list data; the frame store is accessed only when it becomes necessary to resolve ambiguous situations. The major processing steps in joint identification are as follows:

- *Cleaning the hit list:* The expected image is of two straight or curved stripe sections. Isolated hits with no close neighbours are removed, as are hits falling below some minimum threshold strength.

- *Straight-line or curve fitting:* Line or curve segments are fitted to the hit list data corresponding to the two sheets of metal forming the weld joint. Up to now most work has been done using simple flat sheet joints giving straight-line stripe images; only straight-line stripe segments are considered in what follows. Maximum use is made of any *a priori* information available. Initially it is expected that the joint should appear somewhere near the centre of the image. The type and sense of the joint determine the expected stripe angles. After the first frame is processed corresponding to the start of the seam, subsequent joint positions and stripe angles are analysed using the previous data as the starting guess. From the stripe data, the joint angles and plate thicknesses are computed and checked where possible against the information given.

- *End finding:* A closer look is taken at the data from the region around the joint in order to locate the edge of the upper sheet of metal accurately. It is occasionally necessary to use frame store data for this operation.

- *Joint identification:* From a description of the lines defining the metal sheets and the edge of the upper sheet, and knowledge of the joint type, the joint position can be computed. For a lap joint, a line is constructed corresponding to a stripe from a surface perpendicular to the base sheet to intercept the top metal edge. For a simple fillet joint, the intersection of the two stripes defines the joint position. Joint fit-up errors can also be determined, for a lap by computing the distance between the two stripes and knowing the metal thickness, and for the fillet from the distance between the edge of the upper sheet and the joint.

- *Continuity checking:* Frames are analysed at close intervals along the seam. Checks are made on the continuity of the joint positions and stripe angles.

- *Create joint database:* A data record is created defining joint details for the length of each seam. This information will be used at weld time to determine tracking position errors. The same information is also written out to a backing store so that the job can be recalled at any future date without the need for reteaching.

Vision processing during welding

The vision processing situation is far less favourable during welding than it is at teach time. In order for the sensor feedback to be useful, the processing time during the actual welding must be kept to a minimum. The welding process with associated arc glare and flying weld spatter contributes considerably to the image noise. As the components are not specially prepared, it is important to deal with specular reflection. The specular nature of the surfaces, creating both large variations in stripe intensity and multiple reflections, dominates the vision processing problem and can make stripe fitting as done during the vision teach phase very difficult. The shapes of the components being welded are, however, very predictable, the uncertainty being mainly in their relative spatial location. The orientation of the individual stripe images should remain virtually unchanged. This fact allows us to locate the stripe images using a simple correlation technique which is quick to compute and has considerable immunity to image noise.

- *Histogram construction:* For each of the two metal sheets independently, a histogram is constructed of the number of samples against stripe displacement. A first guess of the joint row number is made, which initially will be the teach time value and subsequently will be based on the last computed position. This row number is used to split the range used for the two histograms. For each raster line in the range of interest, the difference between the observed hit column number and the teach time column number is computed and used to build the histogram. The position of the single mode of the histogram indicates the shift in stripe position from the taught value. The dispersion of the histogram is used as a measure of the image quality.

- *Joint identification:* Using the two new stripe definitions the joint position is computed as for the teach time situation described above. If this leads to a joint row value very different from the initial guess, then this new value is used to recompute the histograms.

- *End finding:* As for the teach time case, detailed examination of the joint region is used to locate the edge of the upper plate.

- *Error measurement:* By comparing the measured joint position with the teach time model, an estimate of the lateral and stand-off errors is calculated. The joint gap size is also computed as for the teach time situation described above. These measurements are transmitted back to the control computer in order to close the feedback loop.

Results

The system as described here has been working in the laboratory since March 1983. During this period large numbers of lap and fillet joint samples have been successfully welded. Positional errors in excess of 10mm can be corrected to about 0.5m and gaps detected with an accuracy of about 0.2mm. The system can even cope with the task of correctly tracking the joint while the sample is slowly moved during the actual welding. The design of the sensor has been shown to be capable of survival in the welding environment. Tests have begun with BL Technology Ltd to determine long-term reliability under actual production conditions. Software enhancements are in hand to allow for a greater range of joint configurations. Other work in progress includes the repackaging of the computing system in a form suitable for production and conversions to mount the package on other robots.

References

[1] Clocksin, W.F., Barratt, J.W., Davey, P.G., Morgan, C.G. and Vidler, A.R. 1982. Visually guided robot arc-welding of thin sheet pressings. In, *Proc. 12th Int. Symp. on Industrial Robots*, pp.225-240. IFS (Publications) Ltd, Bedford, UK.

[2] Clocksin, W.F., Davey, P.G., Morgan, C.G. and Vidler, A.R. 1982. Progress in visual feedback for robot arc-welding of thin sheet steel. In, *Proc. 2nd Int. Conf. on Robot Vision and Sensory Controls*, pp.189-200. IFS (Publications) Ltd, Bedford UK.

[3] Agin, G.J. and Binford, T.O. 1973. Computer description of curved objects. In, *Proc. 3rd Int. Conf. on Artificial Intelligence*, Stanford, CA, USA.

[4] Hill, J. and Park, W.T. 1979. Real-time control of a robot with a mobile camera. In, *Proc. 9th Int. Symp. on Industrial Robots*, SME, Dearborn, MI, USA.

[5] Popplestone, R.J. and Ambler, A.P. 1977. Forming body models from range data, Research Report 46, Department of Artificial Intelligence, University of Edinburgh, UK.

[6] Hunter, J.J., Bryce, G.W. and Doherty, J. 1980. On-line control of the arc welding process. In, *Proc. Int. Conf. on Developments in Mechanised, Automated and Robotic Welding,* London, UK.

[7] Masaki, I., Gorman, R.R., Shulman, B.A., Dunne, M.J. and Toda, H. 1981. Arc welding robot with vision. In, *Proc. 11th Int. Symp. on Industrial Robots.* JIRA, Tokyo.

[8] Bamba, T., Maruyama, H., Ohno, E. and Shiga, Y. 1981. A visual sensor for arc welding robots. In, *Proc. 11th Int. Symp. on Industrial Robots.* JIRA, Tokyo.

[9] Linden, G., Lindskog, G. and Nilsson, L. 1980. A control system using optical sensing for metal-inert-gas arc welding. In, *Proc. Int. Conf. on Developments in Mechanised, Automated and Robotic Welding,* London, UK.

Authors' organisations and addresses

A. Agrawal
GCA Corporation
Industrial Systems Group
One Energy Center
Naperville, IL 60566
USA

H.S. Baird
AT & T
Bell Laboratories
600 Mountain Avenue
Murray Hill, NJ 07947
USA

G. Beni
Center for Robotic Systems
  in Microelectronics
University of California
Santa Barbara, CA 93106
USA

R.C. Bolles
SRI International
333 Ravenswood Avenue
Menlo Park, CA 94025
USA

A.J. Cronshaw
PA Technology
Cambridge Laboratory
Melbourn
Royston
Herts SG8 6DP
England

P.A. Fehrenbach
Industrial Systems Division
GEC Research Corporation
Marconi Research Centre
West Hanningfield Road
Great Baddow
Chelmsford
Essex CM2 8HN
England

W.B. Heginbotham
URB and Associates
14 Middleton Crescent
Beeston
Nottingham NG9 2TH
England

D.A. Hill
CUEL
12 Tulip Tree Avenue
Kenilworth
Warwickshire CV8 2BU
England

S.W. Holland
Computer Science Department
General Motors Research
  Laboratories
Warren, MI 48090-9055
USA

C. Loughlin
Electronic Automation Ltd
Haworth House
202 High Street
Hull HU1 1HA
England

M. Lurie
RCA Laboratories
Washington Road
Princeton, NJ 05840
USA

H.W. Mergler
Case Western Reserve University
Cleveland, OH 44106
USA

C.G. Morgan
Meta Machines Ltd
9 Blacklands Way
Abingdon Industrial Park
Abingdon
Oxford OX14 1DY
England

J. Morris
Unimation (Europe) Ltd
Unit C
Stafford Park 18
Telford
England

N. Nimrod
Case Western Reserve University
Cleveland, OH 44106
USA

J.E. Orrock
Honeywell Inc.
Technology Strategy Center
1000 Boone Avenue North
Golden Valley, MN 55427
USA

C.J. Page
Coventry Lanchester Polytechnic
Department of Production
Engineering
Priory Street
Coventry CV1 5FB
England

L.J. Pinson
University of Colorado
Computer Science Department
Colorado Springs, CO 80907
USA

A. Pugh
University of Hull
Department of Electronic
 Engineering
Cottingham Road
Hull HU6 7RX
England

P.P.L. Regtien
Technische Hogeschool Delft
Delft University of Technology
Department of Electrical Engineering
PO Box 5031
2600 GA Delft
The Netherlands

M. Rioux
Electronics Engineering Section
National Research Council of Canada
Montreal Road
Ottawa
Ontario K1A 0RA
Canada

H.E. Schroeder
EG & G Reticon
345 Potrero Avenue
Sunnyvale, CA 94086
USA

D.G. Whitehead
University of Hull
Department of Electronic
Engineering
Hull HU6 7RX
England

R.F. Wolffenbuttel
Technische Hogeschool Delft
Delft University of Technology
Department of Electrical Engineering
PO Box 5031
2600 GA Delft
The Netherlands

Source of material

Robot sensors – A personal view
First presented at the 1985 International Conference on Advanced Robotics, 9-10 September 1985, Tokyo, Japan.
Reprinted courtesy of the author and the Japan Industrial Robot Association.

Robot vision – An evaluation of imaging sensors
First published in the Journal of Robotic Systems, Vol. 1, No. 3, 1984.
Reprinted courtesy of the author and John Wiley & Sons, Inc.

A low-resolution vision sensor
First published in the Journal of Physics E: Sci. Instrum., Vol. 17, 1984.
Reprinted courtesy of the authors and the Institute of Physics.

An integrated vision/range sensor
First presented at the 3rd International Conference on Robot Vision and Sensory Controls, 6-10 November 1983, Cambridge, Massachusetts, USA.
Reprinted courtesy of the authors and IFS (Conferences) Ltd.

Precise robotic assembly using vision in the hand
First presented at the 3rd International Conference on Robot Vision and Sensory Controls, 6-10 November 1983, Cambridge, Massachusetts, USA.
Reprinted courtesy of the authors and IFS (Conferences) Ltd.

Line, edge and contour following with eye-in-hand vision system
First presented at the 14th International Symposium on Industrial Robots, 2-4 October 1984, Gothenburg, Sweden.
Updated in September 1985. Reprinted courtesy of the authors and IFS (Conferences) Ltd.

A novel solid-state colour sensor suitable for robotic applications
First presented at the 5th International Conference on Robot Vision and Sensory Controls, 29-31 October 1985, Amsterdam.
Reprinted courtesy of the authors and IFS (Conferences) Ltd.

Robot eye-in-hand using fibre optics
First presented at the 3rd International Conference on Robot Vision and Sensory Controls, 6-10 November 1983, Cambridge, Massachusetts, USA.
Reprinted courtesy of the authors and IFS (Conferences) Ltd.

Optical alignment of dual-in-line components for assembly

First presented at the 3rd International Conference on Robot Vision and Sensory Controls, 6-10 November 1983, Cambridge, Massachusetts, USA.
Reprinted courtesy of the author and IFS (Conferences) Ltd.

Dynamic sensing for robots – An analysis and implementation

First presented at the 3rd International Conference on Robot Vision and Sensory Controls, 6-10 November 1983, Cambridge, Massachusetts, USA.
Reprinted courtesy of the authors and IFS (Conferences) Ltd.

A practical vision system for use with bowl feeders

First presented at the 1st International Conference on Assembly Automation, 25-27 March 1980, Brighton, UK.
Updated in September 1985. Reprinted courtesy of the authors and IFS (Conferences) Ltd.

A laser-based scanning range-finder for robotic applications

First presented at th 2nd International Conference on Robot Vision and Sensory Controls, 2-4 November 1982, Stuttgart, Germany.
Reprinted courtesy of the authors and IFS (Conferences) Ltd.

Laser range-finder based on synchronised scanners

First published in Applied Optics, Vol. 23, No. 21, 1984.
Reprinted courtesy of the author and the National Research Council of Canada.

The orientation of difficult components for automatic assembly

First presented at the 1st International Conference on Systems Engineering, Coventry Lanchester Polytechnic, September 1980.
Reprinted courtesy of the authors and Coventry Lanchester Polytechnic.

Scene illumination

Not previously published.

CONSIGHT 1 – A vision controlled robot system for transferring parts from belt conveyors

First presented at the GMR Symposium: Computer Vision and Sensor-Based Robots, September 1978, General Motors Research Laboratories, Warren, Michigan, USA.
Reprinted courtesy of the authors, General Motors and Plenum Press.

Practical illumination concept and technique for machine vision applications

First presented at the Robots 8 Conference, 4-7 June 1984, Detroit, Michigan, USA (SME Technical Paper MS84-397).
Reprinted courtesy of the authors, EG & G Reticon and the Society of Manufacturing Engineers.

Three-dimensional locating of industrial parts

First presented at the 8th NSF Grantees Conference on Production Research and Technology, 27-29 January 1981.
Reprinted courtesy of the author, SRI International and the National Science Foundation.

Visual guidance techniques for robot arc-welding
First presented at the 3rd International Conference on Robot Vision and Sensory Controls, 6-10 November 1983, Cambridge, Massachusetts, USA.
Reprinted courtesy of the authors and IFS (Conferences) Ltd.

Contents of Volume 2

1. Passive and Active Force Sensors

2. Tactile Array Sensors

3. Tactile Transducers

30878
10/6/80
IPS.